QGIS Blueprints

Develop analytical location-based web applications
with QGIS

Ben Mearns

open source*
community experience distilled

BIRMINGHAM - MUMBAI

QGIS Blueprints

First published: September 2015

Production reference: 1210915

Published by Packt Publishing Ltd.
Livery Place
35 Livery Street
Birmingham B3 2PB, UK.

ISBN 978-1-78528-907-1

www.packtpub.com

Credits

Author
Ben Mearns

Reviewers
Ujaval Gandhi
Fred Gibbs
Gergely Padányi-Gulyás
Abdelghaffar KHORCHANI
Pablo Pardo
Mats Töpel

Acquisition Editor
Divya Poojari

Content Development Editor
Zeeyan Pinheiro

Technical Editor
Humera Shaikh

Copy Editor
Shruti Iyer

Project Coordinator
Suzanne Coutinho

Proofreader
Safis Editing

Indexer
Monica Ajmera Mehta

Production Coordinator
Nilesh R. Mohite

Cover Work
Nilesh R. Mohite

About the Author

Ben Mearns lives in Philadelphia, PA, where he consults, teaches, advises, speaks, and creates around geographic information. At present, he is involved in private practice; previously, he held the positions of the Lead Geospatial Information Consultant and Instructor of GIS for Natural Resource Management at the University of Delaware. Ben has held other GIS and data positions at the University of Pennsylvania, Cartographic Modeling Lab, Princeton University, and Macalester College. He has consulted in private practice on projects in many fields, including research, healthcare, education, and e-commerce.

I would like to thank Catherine Moore and Raiana Mearns for their support during the writing of this book. I am indebted to Professor John Mackenzie, of the University of Delaware, whose GIS curriculum inspired much of the material in this book. I must also acknowledge others at UD: the caring support of my colleagues in IT research computing and CS&S, the UD GIS community, the lima bean research team, the digital humanities community, and many more.

About the Reviewers

Ujaval Gandhi holds a master's degree in the field of geoinformatics from the University of Wisconsin-Madison, and he has 10 years of experience in the field of remote sensing and GIS. Ujaval is passionate about open source software and an active contributor to the QGIS community. He is currently located in Bangalore, India, and works as a tech manager in the aerial imagery team at Google.

Fred Gibbs is an assistant professor of history at the University of New Mexico, where he teaches the intersection of the history of medicine and urban ecologies, particularly the spatial relationships between people, food, health policy, public health, and urban design. In his research, Fred experiments with new, technology-driven methodologies to create and disseminate historical interpretations. His recent work focuses on creating interactive maps to visualize and understand urban spatial experiences and health. Fred coaches students on the history of medicine, food history, urban health, and digital humanities.

Gergely Padányi-Gulyás is a geographic information systems (GIS) developer, web developer, and remote sensing analyst with over 7 years of experience. He specializes in designing and developing web mapping applications and GIS. Gergely is a dedicated user and developer of open source software (OSS) and also an active member of the Hungary OSGeo chapter. He is familiar with both client- and server-side programming.

For more than 4 years, Gergely worked for archaeologists as a GIS engineer and remote sensing analyst, where he contributed to laying the foundations of archaeological predictive modeling in Hungary. After this, he became a Java web developer in a private company. For the last 2 years, Gergely has been working at a state nonprofit corporation as a lead GIS and web developer, where he uses the experience he gained in his previous jobs: combining GIS with development. In the past few years, he dived into plugin development in different programming languages, such as Java for GeoServer and Python for QGIS. Quite recently, Gergely has been dabbling in WebGIS 3D visualization and analysis.

You can follow him on his personal website at www.gpadanyig.com.

Abdelghaffar KHORCHANI has a license degree in geographic information systems (GIS) and a fundamental license of natural science in applied biology and geology. He holds a master's degree in geomatics and planning, and he is also a computer engineer. Currently, Abdelghaffar is pursuing a master's degree in planning and regional development from Laval University, Canada and a PhD in marine sciences from the University of Milano-Biccoca, Italy.

He organized courses in Japan based on the fisheries resource management approach for young leaders. In Spain, Abdelghaffar designed courses that focus on the field use of GIS for scheduling and management. Besides, he has also devised courses on urban administration in Tunisia.

Abdelghaffar has 8 years of experience in the geomatics field and has worked on several projects in the sectors of agriculture, environment, transport, mapping, and so on. Currently, he manages the Ministry of Agriculture & Environment in Tunisia and is responsible for the mapping service in the vessel monitoring system (VMS) project.

Abdelghaffar is also a trainer in the mapping field of GIS, GPS, and computer-aided design (CAD). He is particularly interested in the development of decision support tools.

I would like to specially thank Packt Publishing for giving me the opportunity of reviewing of this book. I would also like to thank my family, especially my parents, for their physical and moral support.

Pablo Pardo is a geographer from Spain. He has an MSc degree in GIS and specialized in natural risk assessment, focusing his thesis on open data quality. Pablo also has a certificate of higher education in software development.

After several years of working as a GIS technician, he is now beginning his freelance career, combining GIS consulting with data analysis and programming. This is the second book Pablo has helped review.

He likes open data, free software, and "geo stuff". You can find more information about him at www.pablopardo.es.

Mats Töpel received his PhD in systematic biology from the University of Gothenburg, where he studied evolutionary and biogeographical patterns in plants using GIS and climate niche modeling. He is currently working as a bioinformatician specialized in de novo genome sequencing. Mats is the lead programmer in the SpeciesGeoCoder project, a tool for large-scale biogeographical data analysis. When not involved in science, he enjoys spending time with the family or fly-fishing along the coast of Sweden.

www.PacktPub.com

Support files, eBooks, discount offers, and more

For support files and downloads related to your book, please visit www.PacktPub.com.

Did you know that Packt offers eBook versions of every book published, with PDF and ePub files available? You can upgrade to the eBook version at www.PacktPub.com and as a print book customer, you are entitled to a discount on the eBook copy. Get in touch with us at service@packtpub.com for more details.

At www.PacktPub.com, you can also read a collection of free technical articles, sign up for a range of free newsletters and receive exclusive discounts and offers on Packt books and eBooks.

https://www2.packtpub.com/books/subscription/packtlib

Do you need instant solutions to your IT questions? PacktLib is Packt's online digital book library. Here, you can search, access, and read Packt's entire library of books.

Why subscribe?

- Fully searchable across every book published by Packt
- Copy and paste, print, and bookmark content
- On demand and accessible via a web browser

Free access for Packt account holders

If you have an account with Packt at www.PacktPub.com, you can use this to access PacktLib today and view 9 entirely free books. Simply use your login credentials for immediate access.

Table of Contents

Preface

QGIS, the world's most popular free/open source desktop geographic information system software, enables a wide variety of use cases involving location—formerly only available through expensive, specialized commercial software. However, designing and executing a multitiered project from scratch on this complex ecosystem remains a significant challenge.

This book starts with a primer on QGIS and closely-related data, software, and systems. We'll guide you through seven use-case blueprints for geographic web applications. Each blueprint boils down a complex workflow into steps that you can follow to reduce the time usually lost to trial and error.

By the end of this book, readers will be able to build complex layered applications that visualize multiple datasets, employ different types of visualization, and give end users the ability to interact with and manipulate this data for the purpose of analysis.

What this book covers

Chapter 1, *Exploring Places — from Concept to Interface*, gives you an overview of the application types and technical aspects that will be covered and how QGIS will be leveraged. through a simple map application example from the digital humanities. You will use some fundamental GIS techniques to produce a tile cache-leveraging web application that includes geocoded addresses, joined data, and a georeferenced image.

Chapter 2, *Identifying the Best Places*, will look at how raster data can be analyzed, enhanced, and used for map production. You will produce suitability grids using map algebra and export to produce a simple click-based map.

Chapter 3, *Discovering Physical Relationships*, will create an application for physical raster modeling. We will use raster analysis and a model automation tool to model the physical conditions for some basic hydrological analysis. Finally, you will use a cloud platform to enable dynamic query from the client-side application code.

Chapter 4, Finding the Best Way to Get There, will explore formal, network-like geographic relationships between vector objects. You will learn to create a few visualizations related to optimal paths: isochron polygons and accumulated traffic lines. The end result will be an application that will communicate back and forth with your audience regarding safe school routes.

Chapter 5, Demonstrating Change, will demonstrate visualization and analytical techniques to explore relationships between place and time and between places themselves. You will work with demographic data from a census for election purposes through a timeline controlled animation.

Chapter 6, Estimating Unknown Values, will use interpolation methods to estimate unknown values at one location based on the known values at other locations using the NetCDF array-oriented scientific data format. You will use parameters such as precipitation, relative humidity, and temperature to predict the vulnerability of fields and crops to mildew.

Chapter 7, Mapping for Enterprises and Communities, will use a mix of web services to provide a collaborative data system. You will create an editable and data-rich map for the discovery of community information by accessing information about community assets.

What you need for this book

You will need:

- QGIS 2.10
- A computer running OS X, Windows, or Linux

Who this book is for

This book is for relatively experienced GIS developers who have a strong grounding in the fundamentals of GIS development. They must have used QGIS before, but are looking to understand how to develop more complex, layered map applications that effectively expose various datasets and visualizations.

Conventions

In this book, you will find a number of text styles that distinguish between different kinds of information. Here are some examples of these styles and an explanation of their meaning.

Code words in text, database table names, folder names, filenames, file extensions, pathnames, dummy URLs, user input, and Twitter handles are shown as follows: "Run QTiles, creating a new `mytiles` tileset with a minimum zoom of `14` and maximum of `16`."

A block of code is set as follows:

```
[..] > cd c:\packt\c4\data\output
c:\packt\c4\data\output>java -jar osm2po-5.0.0\osm2po-core-5.0.0-
  signed.jar cmd=tj
sp newark_osm.osm
```

Any command-line input or output is written as follows:

```
cd C:\packt\c6\data\web
python -m CGIHTTPServer 8000
```

New terms and **important words** are shown in bold. Words that you see on the screen, for example, in menus or dialog boxes, appear in the text like this: "Navigate to **Layer | Open attribute table**."

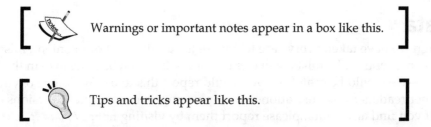

> Warnings or important notes appear in a box like this.

> Tips and tricks appear like this.

Reader feedback

Feedback from our readers is always welcome. Let us know what you think about this book — what you liked or disliked. Reader feedback is important for us as it helps us develop titles that you will really get the most out of.

To send us general feedback, simply e-mail feedback@packtpub.com, and mention the book's title in the subject of your message.

If there is a topic that you have expertise in and you are interested in either writing or contributing to a book, see our author guide at www.packtpub.com/authors.

Customer support

Now that you are the proud owner of a Packt book, we have a number of things to help you to get the most from your purchase.

Downloading the example code

You can download the example code files from your account at `http://www.packtpub.com` for all the Packt Publishing books you have purchased. If you purchased this book elsewhere, you can visit `http://www.packtpub.com/support` and register to have the files e-mailed directly to you.

Downloading the color images of this book

We also provide you with a PDF file that has color images of the screenshots/diagrams used in this book. The color images will help you better understand the changes in the output. You can download this file from: `https://www.packtpub.com/sites/default/files/downloads/9071OS_Graphics.pdf`.

Errata

Although we have taken every care to ensure the accuracy of our content, mistakes do happen. If you find a mistake in one of our books—maybe a mistake in the text or the code—we would be grateful if you could report this to us. By doing so, you can save other readers from frustration and help us improve subsequent versions of this book. If you find any errata, please report them by visiting `http://www.packtpub.com/submit-errata`, selecting your book, clicking on the **Errata Submission Form** link, and entering the details of your errata. Once your errata are verified, your submission will be accepted and the errata will be uploaded to our website or added to any list of existing errata under the Errata section of that title.

To view the previously submitted errata, go to `https://www.packtpub.com/books/content/support` and enter the name of the book in the search field. The required information will appear under the **Errata** section.

Piracy

Piracy of copyrighted material on the Internet is an ongoing problem across all media. At Packt, we take the protection of our copyright and licenses very seriously. If you come across any illegal copies of our works in any form on the Internet, please provide us with the location address or website name immediately so that we can pursue a remedy.

Please contact us at copyright@packtpub.com with a link to the suspected pirated material.

We appreciate your help in protecting our authors and our ability to bring you valuable content.

Questions

If you have a problem with any aspect of this book, you can contact us at questions@packtpub.com, and we will do our best to address the problem.

1
Exploring Places – from Concept to Interface

How do we turn our idea into a location-based web application? If you've heard this question before or asked it yourself, you would know that this deceptively simple question can have answers posed in a limitless number of ways. In this book, we will consider the application of QGIS through specific use cases selected for their general applicability. There's a good chance that the blueprint given here will shed some light on this question and its solution for your application.

In this book, you will learn how to leverage this ecosystem, let the existing software do the heavy lifting, and build the web mapping application that serves your needs. When integrated software is seamlessly available in QGIS, it's great! When it isn't, we'll look at how to pull it in.

In this chapter, we will look at how data can be acquired from a variety of sources and formats and visualized through QGIS. We will focus on the creation of the part of our application that is relatively static: the basemap. We will use the data focused on a US city, Newark, Delaware. A collection of data, such as historical temperature by area, point data by address, and historical map images, could be used for a digital humanities project, for example, if one wanted to look at the historical evidence for lower temperatures observed in a certain part of a city.

In this chapter, we will cover the following topics:

- The software
- Extract, Transfer, and Load
- Georeference
- The table join
- Geocoding

- Orthorectification
- The spatial reference manipulation
- The spatial reference assignment
- Projection
- Transformation
- The basemap creation and configuration
- Layer scale dependency
- Labeling
- The tile creation

The software

QGIS is not a black box—it is a part of a dynamic community of software users and developers. Although this book strives to apply generalizable blueprints to a range of actual web applications, there are times when a deeper understanding of the makeup of the QGIS platform is invaluable; sometimes even as soon as at installation. The latest version of QGIS is 2.10.

The development community and dependencies

As QGIS is open source, no one entity owns the project; it's supported by a well-established community. The project is guided by the QGIS **Project Steering Committee** (**PSC**), which selects managers to oversee various areas of development, testing, packaging, and other infrastructure to keep the project going. The **Open Source Geospatial Foundation** (**OSGeo**) is a major contributor to software development, and QGIS is considered an official OSGeo project. Many of QGIS' dependencies and complimentary software are also OSGeo projects, and this collective status has served to bring some integration into what can be considered a platform. The **Open GIS Consortium** (**OGC**) deliberates and sets standards for the data and metadata formats. QGIS supports a range of OGC standards—from web services to data formats.

When QGIS is at its best, this rich platform provides a seamless functionality, with an ecosystem of open or simply available software ready to be tapped. At other times, the underlying dependencies and ecosystem software require more attention. Since it's an open source software, contributions are always being made, and you have the option of making customizations in code and even contributing to it!

Data format read/write

The OSGeo ecosystem provides capabilities for data format read/write through the **OGR Simple Features Library** (OGR, originally for OpenGIS Simple Features Reference Implementation) and **Geospatial Data Abstraction Library (GDAL)** libraries, which support around 220 formats.

Geospatial coordinate transformation

The models of the earth, which the coordinates refer to, are collectively known as **Coordinate Reference Systems (CRSs)**. The spatial reference transformation between systems and projection—from a system in linear versus the one in angular coordinates—is supported by the PROJ.4 library with around 2,700 systems. These are expressed in a plain text format defined by PROJ.4 as **Well Known Text (WKT)**. PROJ.4 WKT is actually very readable, containing the sort of information that would be familiar to the students of cartographic projection, such as meridians, spheroids, and so on.

Analysis

Analysis, or application of algorithmic functions to data is rarely handled seamlessly by QGIS. More often, it is an extension of one of the dependencies already listed before or is provided by **System for Automated Geoscientific Analyses (SAGA)**. Many other analytical operations are provided by numerous QGIS Python plugins.

In general, these libraries will seamlessly transform to or from the formats that we require. However, in some cases, additional dependencies will need to be acquired and either be built and configured themselves or have the code built around them.

Web publishing

QGIS has the capability of publishing to web hosts through both integrated and less immediate means.

Installation

OSGeo project binaries have sometimes been bundled to ease the installation process, given the multitude of interdependencies among projects. Tutorials in this book are written based on an installation using the QGIS standalone installer for Windows.

Linux

QGIS hosts repositories with the most current versions for Debian/Ubuntu and bundled packages for other major Linux distributions; however, these repositories are generally many versions behind. You will find that this is often the case even with the extra repositories for your distribution (for example, EPEL for RHEL flavors). Seeking out other repositories is worthwhile. Another option, of course, is to attempt to build it from scratch; however, this can be very difficult.

Mac

There is no bundled package installer for Mac OS, though you should be able to install QGIS with only one or two additional installations from the binaries readily available on the Web—the KyngChaos Wiki has long been the go-to source for this.

Windows

Installation with Windows is simpler than with other platforms at this time. The most recent version of QGIS, with basic dependencies such as GDAL, is installed with a typical executable installer: the "standalone" installer. In addition, the OSGeo4W (OSGeo for Windows) package installer is very useful for the extended dependencies. You will likely find that beyond simply installing QGIS, you will return to this installer to add additional software to extend QGIS into its ecosystem. You can launch the installer from the **Setup** shortcut under the QGIS submenu in the Windows Start menu.

OSGeo-Live

The most extensive incarnation of the OSGeo software is embodied in OSGeo-Live, a Lubuntu **Virtual Machine (VM)** on which all of the OSGeo software is already installed. It is listed here separately since it will boot into its own OS, independent of the host platform.

Updates to OSGeo Live are typically released in tandem with FOSS4G, an annual global event hosted by OSGeo since 2006. Given that these events occur less regularly and are out of sync with OSGeo software development, bundled versions are usually a few releases behind. Still, OSGeo-Live is a quick way to get started.

Now that you've prepared your local machine, let's return to the idea of the generalizable web applications that will be the focus of this book. There are a few elements that we can identify in the process of developing web-mapping applications.

Acquiring data for geospatial applications

After any preliminary planning — a step that should include careful consideration of at least the use cases for our application — we must acquire data. Acquisition involves not only the physical transfer of the data, but also processing the data to a particular format and importing it into whatever data storage scheme we have developed. This is usually called **Extract, Transform, and Load** (ETL).

Though ETL is the first major step in developing a web application, it should not be taken lightly. As with any information-based project, data often comes to us in a form that's not immediately useable — whether because of nonuniform formatting, uncertain metadata, or unknown field mapping. Although any of these can affect a GIS project, as GISs are organized around cartographic coordinate systems, the principle concern is usually that data must be spatially described in a uniform way, namely by a single CRS, as referred to earlier. To that end, data often requires georeferencing and spatial reference manipulation.

For certain datasets, an ETL workflow is unnecessary because the data is already provided via web services. Using hosted data stored on the remote server and read directly from the Web by your application is a very attractive option, purely for ease of development if nothing else. However, you'll probably need to change the CRS, and possibly other formatting, of your local data to match that of the hosted data since hosted services are rarely provided in multiple CRSs. You must also consider whether the hosted data provides capabilities that support the interface of your application. You will find more information on this topic under the operational layer section of this chapter.

Producing geospatial data with georeferencing

By georeferencing, or attaching our data to coordinates, we assert the geographic location of each object in our data. Once our data is georeferenced, we can call it geospatial. Georeferencing is done according to the fields in the data and those available in some geospatial reference source.

The simplest example is when a data field actually matches a field in some existing geospatial data. This data field is often an ID number or name. This kind of georeferencing is called a **table join**.

Table join

In this example, we will take a look at a table join with some temperature data from an unknown source and census tract boundaries from the US Census. Census' TIGER/Line files are generally the first places to look for U.S. national boundary files of all sorts, not just census tabulation areas.

The temperature data to be georeferenced through a table join would be as follows:

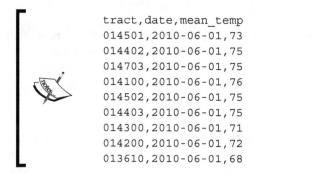

```
tract,date,mean_temp
014501,2010-06-01,73
014402,2010-06-01,75
014703,2010-06-01,75
014100,2010-06-01,76
014502,2010-06-01,75
014403,2010-06-01,75
014300,2010-06-01,71
014200,2010-06-01,72
013610,2010-06-01,68
```

Temperature data metadata would be as follows:

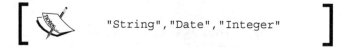

```
"String","Date","Integer"
```

> ### Downloading the example code
>
> You can download the example code files for all Packt books you have purchased from your account at http://www.packtpub.com. If you purchased this book elsewhere, you can visit http://www.packtpub.com/support and register to have the files e-mailed directly to you.

To perform a table join, perform the following steps:

1. Copy the code from the first information box calls into a text file and save this as temperature.csv.

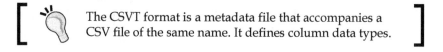

> The CSVT format is a metadata file that accompanies a CSV file of the same name. It defines column data types.

2. Copy the code from the second information box into a text file and save this as `temperature.csvt`. Otherwise, QGIS will not know what type of data is contained in each column.

> Data for all the chapters will be found under the data directory for each chapter. You can use the included data under `c1/data/original` with the file names given earlier. Besides selecting the browse menu, you can also just drag the file into the **Layers** panel from an open operating system window. You can find examples of data output during exercises under the output directory of each chapter's data directory. This is also the directory given in the instructions as the destination directory for your output. You will probably want to create a new directory for your output and save your data there so as to not overwrite the included reference data.

3. Navigate to **Layer | Add Layer | Add Vector Layer | Browse to**, and select `temperature.csv`.

> CSV data can also be added through **Layer | Add Layer | Add Delimited Text**. This is especially useful to plot coordinates in a CSV, as you'll see later.

4. Download the **Tract** boundary data:

    1. Visit `http://www.census.gov/geo/maps-data/data/tiger-line.html`.

    2. Click on the tab for the year you wish to find.

    3. Download the web interface.

    4. This will take us to `http://www.census.gov/cgi-bin/geo/shapefiles2014/main`.

    5. Navigate to **Layer Type | Census Tracts** and click on the **submit** button. Now, select **Delaware** from the **Census Tract (2010)** dropdown. Click on **Submit** again. Now select **All counties in one state-based file** from the dropdown displayed on this page and finally click on **Download**.

    6. Unzip the downloaded folder.

5. Navigate to **Layer | Add Layer | Add Vector Layer | Browse to**, and select the `tl_2010_10_tract10.shp` file in the unzipped directory.

6. Right-click on `tl_2010_10_tract10` in the **Layer** panel, and then navigate to **Properties | Joins**. Click on the button with the green plus sign (**+**) to add a join.

7. Select **temperature** as the **Join layer** option, **tract** as the **Join field** option, **TRACTCE10** as the **Target field** option, and click on **OK** on this and the properties dialog:

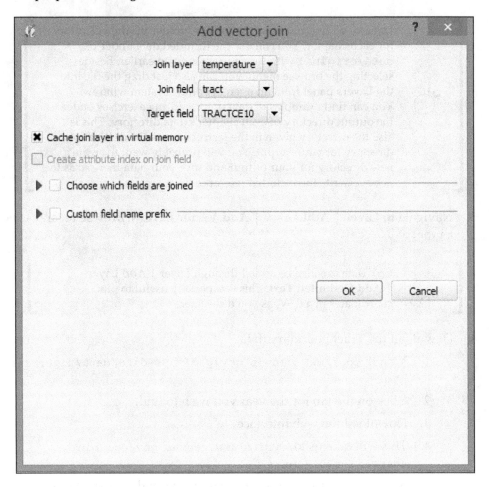

To verify that the join completed, open the attribute table of the target layer (such as the geospatial reference, in this case, t1_2010_10) and sort by the new temperature_mean_temp field. Notice that the fields and values from the join layer are now included in the target layer.

1. Select the target layer, t1_2010_10_tract10, from the **Layers** panel.

2. Navigate to **Layer | Open attribute table**.

3. Click on the temperature_mean_temp column header to sort tracts by this column. You may have to click twice to toggle the sort order from ascending to descending.

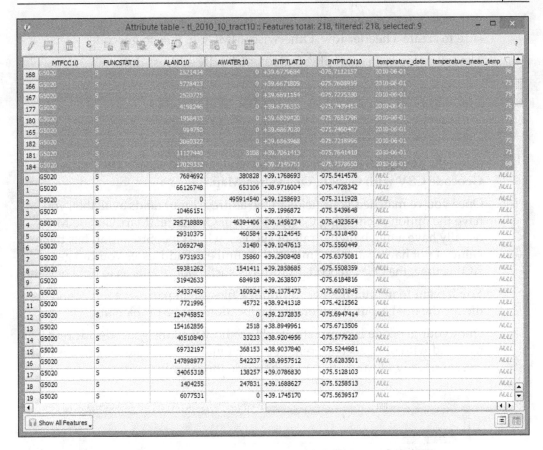

Attribute table - tl_2010_10_tract10 :: Features total: 218, filtered: 218, selected: 9

	MTFCC10	FUNCSTAT10	ALAND10	AWATER10	INTPTLAT10	INTPTLON10	temperature_date	temperature_mean_temp
168	G5020	S	2821434	0	+39.6779684	-075.7112157	2010-06-01	76
166	G5020	S	5728423	0	+39.6671809	-075.7608959	2010-06-01	75
167	G5020	S	2520725	0	+39.6691154	-075.7275330	2010-06-01	75
177	G5020	S	4198246	0	+39.6726333	-075.7439453	2010-06-01	75
180	G5020	S	1958433	0	+39.6809420	-075.7683796	2010-06-01	75
165	G5020	S	994750	0	+39.6867020	-075.7460407	2010-06-01	73
182	G5020	S	3060322	0	+39.6863968	-075.7218996	2010-06-01	72
181	G5020	S	11127440	3108	+39.7061413	-075.7641410	2010-06-01	71
184	G5020	S	17029332	0	+39.7145751	-075.7378650	2010-06-01	68
0	G5020	S	7684692	380828	+39.1768693	-075.5414576	NULL	NULL
1	G5020	S	66126748	653106	+38.9716004	-075.4728342	NULL	NULL
2	G5020	S	0	495914540	+39.1258693	-075.3111928	NULL	NULL
3	G5020	S	10466151	0	+39.1996872	-075.5439648	NULL	NULL
4	G5020	S	295718889	46394406	+39.1456274	-075.4323654	NULL	NULL
5	G5020	S	29310375	460584	+39.2124545	-075.5318450	NULL	NULL
6	G5020	S	10692748	31480	+39.1047613	-075.5560449	NULL	NULL
7	G5020	S	9731933	35860	+39.2908408	-075.6375081	NULL	NULL
8	G5020	S	59381262	1541411	+39.2858685	-075.5508359	NULL	NULL
9	G5020	S	31942633	684918	+39.2638507	-075.6184816	NULL	NULL
10	G5020	S	34337450	160924	+39.1375473	-075.6031845	NULL	NULL
11	G5020	S	7721996	45732	+38.9241318	-075.4212562	NULL	NULL
12	G5020	S	124745852	0	+39.2372835	-075.6947414	NULL	NULL
13	G5020	S	154162856	2518	+38.8949961	-075.6713506	NULL	NULL
14	G5020	S	40510840	33233	+38.9204956	-075.5779220	NULL	NULL
15	G5020	S	69732197	368153	+38.9037840	-075.5244981	NULL	NULL
16	G5020	S	147898977	542237	+38.9957512	-075.6283501	NULL	NULL
17	G5020	S	34065318	138257	+39.0786830	-075.5128103	NULL	NULL
18	G5020	S	1404255	247831	+39.1688627	-075.5258513	NULL	NULL
19	G5020	S	6077531	0	+39.1745170	-075.5639517	NULL	NULL

Show All Features

Geocode

If our data is expressed as addresses, intersections, or other well-known places, we can geocode it (that is, match it with coordinates) with a local or remote geocoder configured for our particular set of fields, such as the standard fields in an address.

In this example, we will geocode it using the remote geocoder provided by Google. Perform the following steps:

1. Install the MMQGIS plugin.
2. If you don't already have some address data to work with, you can make up a delimited file that contains some standard address fields, such as street, city, state, and county (ZIP code is not used by this plugin). The data that I'm using comes from New Castle County, Delaware's GIS site (http://gis.nccde.org/gis_viewer/).

3. Whether you've downloaded your address data or made up your own, make sure to create a header row. Otherwise, MMQGIS fails to geocode.

The following is an example of MMQGIS-friendly address data:

```
id,address,city,state,zip,country
1801300170,44 W CLEVELAND AV,NEWARK,DE,19711,USA
1801400004,85 N COLLEGE AV,NEWARK,DE,19711,USA
1802600068,501 ACADEMY ST,NEWARK,DE,19716,USA
```

4. Open the MMQGIS geocode dialog by navigating to **MMQGIS | Geocode | Geocode CSV with Google/OpenStreetMap**.

5. Once you've matched your fields to the address input fields available, you have the option of choosing Google Maps or OpenStreetMap. Google Maps usually have a much higher rate of success, while OpenStreetMap has the value of not having a daily limit on the number of addresses you can geocode. At this time, the OSM geocoder produces such poor results as to not be useful.

6. You'll want to manually select or input a filesystem path for a notfound.csv file for the final input. The default file location can be problematic.

7. Once your geocode is complete, you'll see how well the geocode address text matched with our geocoder reference. You may wish to alter addresses in the notfound.csv file and attempt to geocode these again.

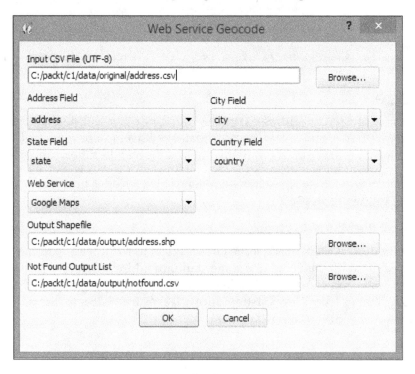

Orthorectify

Finally, if our data is an image or grid (raster), we can match up locations in the image with known locations in a reference map. The registration of these pairs and subsequent transformation of the grid is called **orthorectification** or sometimes by the more generic term, georeferencing (even though that applies to a wider range of operations).

1. Add a basemap, to be used for reference:
    1. Add the OpenLayers plugin. Navigate to **Plugins | Manage | Install Plugins**; select **OpenLayers Plugin** and click on **Install**.
    2. Navigate to **Web | OpenLayers plugin**, and select the basemap of your choice. MapQuest-OSM is a good option.

2. Obtain map image:
    1. I have downloaded a high-resolution image (`c1/data/original/4622009.jpg`) from David Rumsey Map Collection, MapRank Search (`http://rumsey.mapranksearch.com/`), which is an excellent source for historical map images of the United States.
    2. Search by a location, filtering by time, scale, and other attributes. You can find the image we use by searching for Newark, Delaware.
    3. Once you find your map, navigate to it. Then, find **Export** in the upper right-hand corner, and export an extra high-resolution image.
    4. Unzip the downloaded folder.

3. Orthorectify/georeference the image with the following steps:
    1. Install and enable the Georeferencer GDAL plugin.
    2. Navigate to **Raster | Georeferencer | Georeferencer**.
    3. Pan and zoom the reference basemap in the canvas on a location that you recognize in the map image.
    4. Pan and zoom on the map image.
    5. Select **Add Control Point** if it is not already selected.
    6. Click on the location in the map image that you recognized in the third step.
    7. Click the **Pencil** icon to choose control point from Map Canvas.
    8. Click on the location in the reference basemap.
    9. Click on **OK**.

10. Add three of these control points, as shown in the following screenshot:

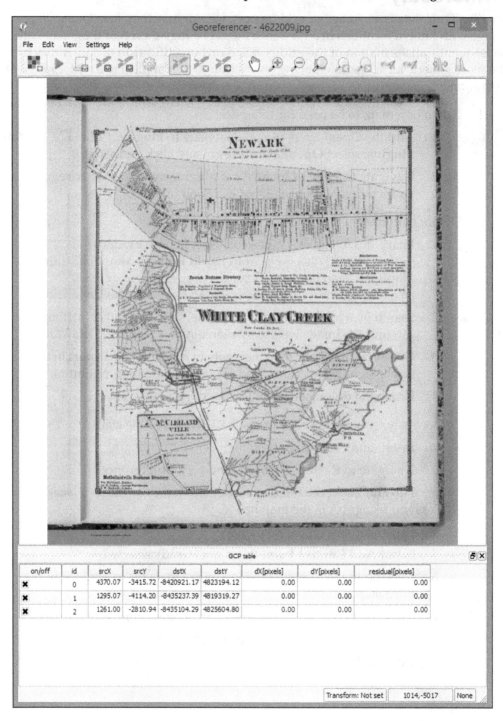

on/off	id	srcX	srcY	dstX	dstY	dX[pixels]	dY[pixels]	residual[pixels]
✖	0	4370.07	-3415.72	-8420921.17	4823194.12	0.00	0.00	0.00
✖	1	1295.07	-4114.20	-8435237.39	4819319.27	0.00	0.00	0.00
✖	2	1261.00	-2810.94	-8435104.29	4825604.80	0.00	0.00	0.00

11. Start georeferencing by clicking on the **Play** button.

12. Enter the transformation settings information, as shown in the following screenshot:

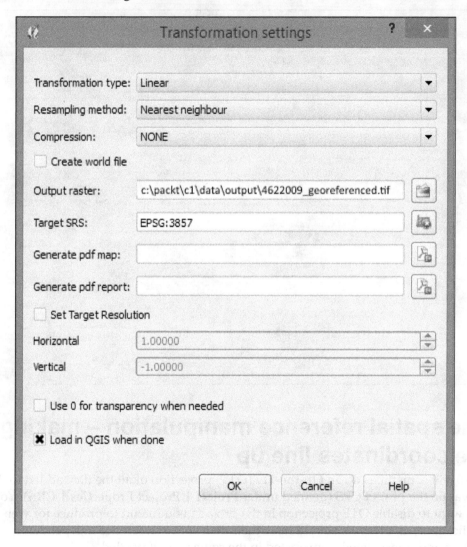

4. Now, start georeferencing by clicking on the **Play** button again.

Once your image has been georeferenced, you should see it align with the other data on your map. You can alter the layer transparency under **Layer properties | Transparency**:

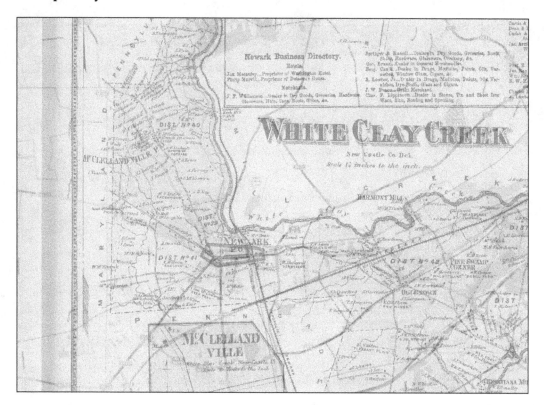

The spatial reference manipulation – making the coordinates line up

QGIS will sometimes do an **On-the-Fly** (OTF) projection of all the data added to the canvas on the project CRS (defined under **Project | Project Properties | CRS**). You will want to disable OTF projection in the projects you intend to produce for web applications, as all layers should have their own spatial reference independently defined and transformed or projected in the same CRS, if needed.

Setting CRS

When geospatial data is received with no metadata on what the spatial reference system describes its coordinates, it is necessary to assign a system. This can be by right-clicking on the layer in **Layers Panel** | **Save as** and selecting the new CRS.

Transformation and projection

At other times, data is received with a different CRS than in the case of the other data used in the project. When CRSs differ, care should be taken to see whether to alter the CRS of the new nonconforming data or of the existing data. Of course we want to choose a system that supports our needs for accuracy or extent; at other times when we already have a suitable basemap, we will want operational layers to conform to the basemap's system. When a suitable basemap is already available to be consumed by our web application, we can often use the system of the basemap for the project. All major third-party basemap providers use **Web Mercator**, which is now known as **EPSG:3857**.

You can project data from geographic to projected coordinates or from one projection to another. This can be done in the same way as you would define a projection: by right-clicking on a layer in **Layers Panel | Save as** and selecting the new CRS. An appropriate transformation will generally be applied by default.

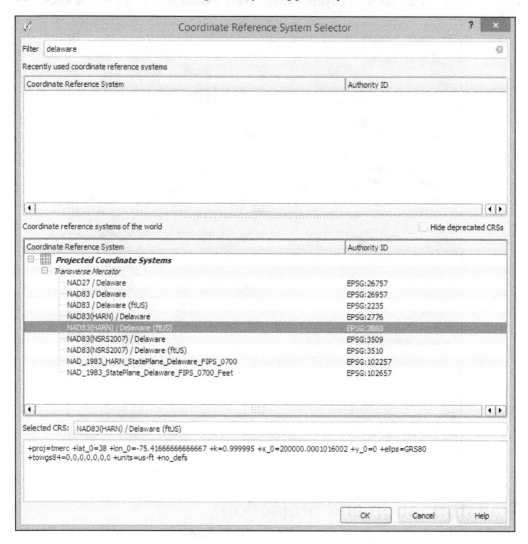

There are some features in CRS Selector that you should be aware of. By selecting from **Recently used coordinate reference systems**, you can often easily match up a new CRS with those existing in the workspace. You also have the option to search through the available systems by entering the **Filter** input. You will see the PROJ.4 WKT representation of the selected CRS at the bottom of the dialog.

Visualizing GIS data

Although the data has been added to the GIS through the ETL process, it is of limited value without adding some visualization enhancements.

The layer style

The layer style is configured through the **Style** tab in the **Layer Properties** dialog. The **Single Symbol** style is the default style type, and it simply shows all the geographic layer objects using the same basic symbol. This setting doesn't provide any visual differentiation between objects other than their apparent geospatial characteristics. The **Categorized** and **Graduated** style types provide different styles according to the attribute table field of your choosing. **Graduated**, which applies to quantitative data, is particularly powerful in the way the color and symbols size are mapped to a numerical scale. This is all accomplished through the **Layer Properties | Style** tab.

To configure a simple graduated layer style to the data, perform the following steps:

1. Under the **Layer Properties | Style** tab, select **Graduated** for the style type.

2. Select your quantitative field for **Column** (such as temperature_mean_temp).

3. Click on **Classify** to group your data into the number of classes specified (by default, **5**) and to select the classification mode (**Equal Interval**, by default):

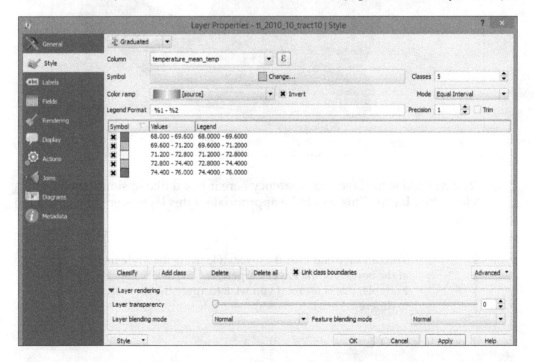

4. Now, click on **Apply**.

    If you applied the preceding steps to the joined tract/temperature layer, you'd see something similar to the following image:

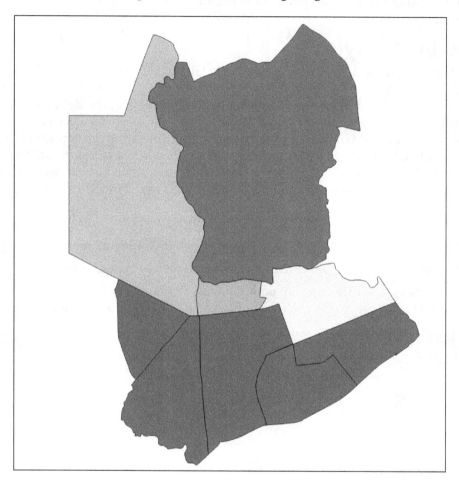

5. You can add some layer transparency here if you'd like to simultaneously view other layers. This would be appropriate if this layer were to be included in a basemap.

You can save and load the layer style using the **Style** menu at the bottom of the **Layer Properties | Style** tab. For example, this is useful if you wish to apply the same style to different layers.

6. Now, click on **OK**.

You will now see the following output:

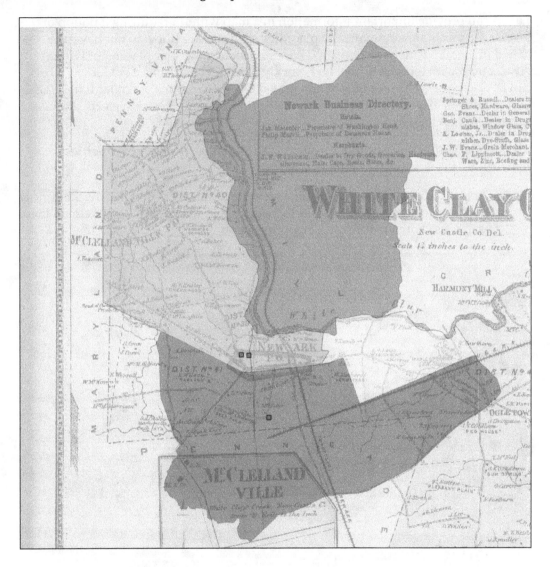

Perhaps the lower temperatures to the north of the city are related to the historical development in other parts of the city.

Labels

Labels can provide important information in a map without requiring a popup or legend. As labels are automatically placed by the software (they do not have an actual physical position in space), they are subject to particular placement issues. Also, a map with too much label text quickly becomes confusing. Sometimes, labels can be easily rendered into tiles by integrated QGIS operations. At other times, this will require an external rendering engine or a map server. These issues are discussed later on in this chapter.

To label the address points in our example, go to that layer's **Layer Properties** | **Labels** tab. Perform the following steps:

1. Select **address** for the **Label this layer with** field. Note that the checkbox next to this selection will be toggled. All other label style options will remain the same, as shown in the following screenshot:

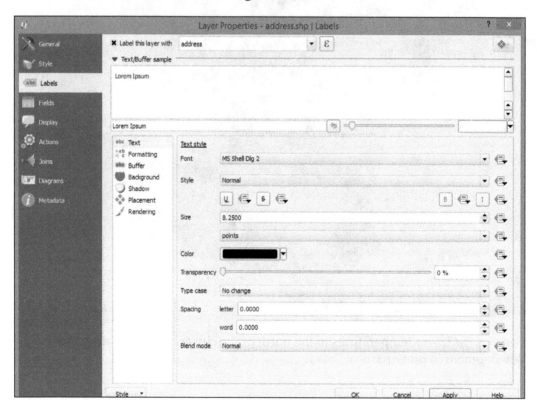

2. Open the **Buffer** subtab and toggle **Draw text buffer**:

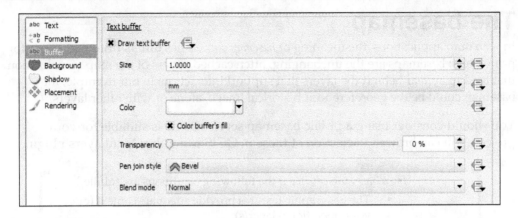

3. After you click on **OK**, you will see something similar to the following screenshot:

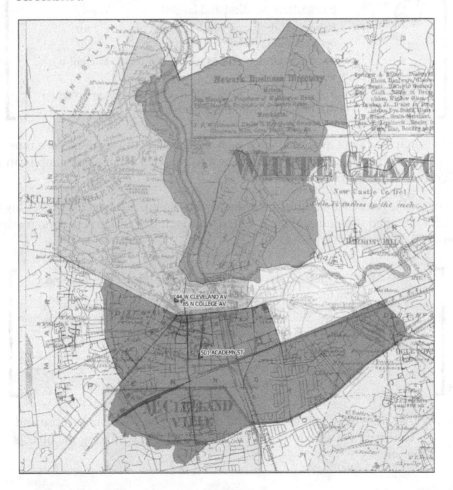

The basemap

In web map applications, the meaning of *basemap* sometimes differs from that in use for print maps; basemaps are the unchanging, often cached, layers of the map visualization that tend to appear behind the layers that support interaction. In our example, the basemap could be the georeferenced historical image, alone or with other layers.

You should consider using a public basemap service if one is suitable for your project. You can browse a selection of these in QGIS using the OpenLayers plugin.

Use a basemap service if the following conditions are fulfilled:

- The geographic features provide adequate context for your operational layer(s)
- The extent is suitable for your map interface
- Scale levels are suitable for your map interface, particularly smallest and largest
- Basemap labels and symbols don't obscure your operational layer(s)
- The map service provides terms of use consistent with your intended use
- You do not need to be able to view the basemap when disconnected from the internet

If our basemap were not available via a web service as in our example, we must turn our attention to its production. It is important to consider what a basemap is and how it differs from the operational layer.

The geographic reference features included in a basemap are selected according to the map's intended use and audience. Often, this includes certain borders, roads, topography, hydrography, and so on.

Beyond these reference features, include the geographic object class in the basemap if you do not need:

- To regularly update the geometric data
- To provide capabilities for style changes
- To permit visibility change in class objects independently of other data
- To expose objects in the class to interface controls

Assuming that we will be using some kind of caching mechanism to store and deliver the basemap, we will optimize performance by maximizing the objects included therein.

Using OpenStreetMap for the basemap data

Obtaining data for a basemap is not a trivial task. If a suitable map service is available via a web service, it would ease the task considerably. Otherwise, you must obtain supporting data from your local system and render this to a suitable cartographic format.

A challenge in creating basemaps and keeping them updated is interacting with different data providers. Different organizations tend be recognized as the provider of choice for the different classes of geographic objects. With different organizations in the mix, different data format conventions are bound to occur.

OpenStreetMap (OSM), an open data repository for geographic reference data, provides both map services and data. In addition to OSM's own map services, the data repository is a source for a number of other projects offering free basemap services.

OpenStreetMap uses a more abstract and scalable schema than most data providers. The OSM data includes a few system fields, such as `osm_id`, `user_id`, `osm_version`, and `way`. The `osm_id` field is unique to each geographic object, `user_id` is unique to the user who last modified the object, `osm_version` is unique to the versions for the object, and `way` is the geometry of the object.

By allowing a theoretically unlimited number of key value pairs along with the system fields mentioned before, the OSM data schema can potentially allow any kind of data and still maintain sanity. Keys are whatever the data editors add to the data that they upload to the repository. The well-established keys are documented on the OSM site and are compatible with the community produced rendering styles. If a community produced style does not include the key that you need or the one that you created, you can simply add it into your own rendering style. Columns are kept from overwhelming a local database during the import stage. Only keys added in a local configuration file are added to the database schema and populated.

High quality cartography with OSM data is an ongoing challenge. CloudMade has created its business on a cloud-based, albeit limited, rendering editor for OSM data, which is capable of also serving map services. CloudMade is, in fact, a fine source for cloud services for OSM data and has many visually appealing styles available. OpenMapSurfer, produced by a research group at the University of Heidelberg, shows off some best practices in high quality cartography with OSM data including sophisticated label placement, object-level scale dependency, careful color selection, and shaded topographic relief and bathymetry.

To obtain the OpenStreetMap data locally to produce your own basemap, perform the following steps:

1. Install the OpenLayers and OSMDownloader QGIS plugins if they are not already installed.

2. Create a new SpatiaLite database.

3. Turn on OSM:

    1. Navigate to **Web | OpenLayers | OpenStreetMap | OpenStreetMap**.

4. Browse your area of interest.

5. Download your area of interest:

    1. Navigate to **Vector | OpenStreetMap | Download Data**:

6. Import the downloaded XML data into a topological SQLite database. This does not contain SpatiaLite geographic objects; rather, it is expressed in terms of topological relationships between objects in a table. Topological relationships are explored in more depth in *Chapter 4, Finding the Best Way to Get There*, and *Chapter 5, Demonstrating Change*.

    1. Navigate to **Vector | OpenStreetMap | Import Topology** from XML.

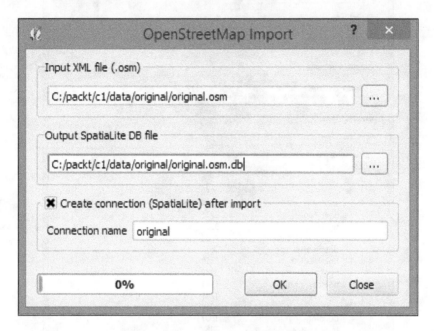

7. Convert topology to SpatiaLite spatial tables through the following steps:

    1. Navigate to **Vector | OpenStreetMap | Export Topology to Spatialite**.

    2. Select the points, polylines, or polygons to export.

    3. Then, select the fields that you may want to use for styling purposes. You can populate a list of possible fields by clicking on **Load from DB**.

4. You can repeat this step to export the additional geometry types, as shown in the following screenshot:

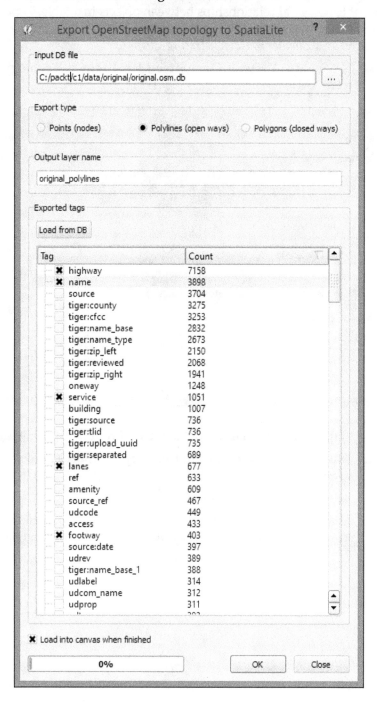

8. You can now style this as you like and export it as the tiled basemap. Then, you can save it in the `mapnik` or `sld` style for use in rendering in an external tile caching software.

Here's an example of the OSM data overlaid on our other layers with a basic, single symbol style:

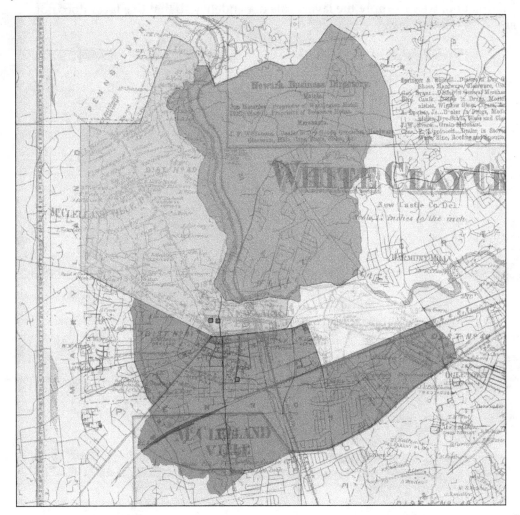

Avoiding obscurity and confusion

Of course, heaping data in the basemap is not without its drawbacks. Other than the relative loss of functionality, which occurs by design, basemaps can quickly become cluttered and otherwise unclear. The layer and label scale dependency dynamically alter the display of information to avoid the obfuscation of basemap geographic classes.

The layer scale dependency

When classes of geographic objects are unnecessary to visualize at certain scales, the whole layer scale dependency can be used to hide the layer from view. For example, in the preceding image, we can see all the layers, including the geocoded addresses, at a smaller scale even when they may not be distinctly visible. To simplify the information, we can apply the layer scale dependency so that this layer does not show these small scales.

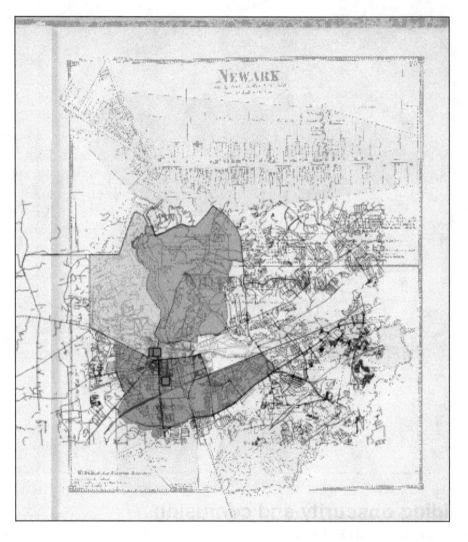

At this scale, some objects are not distinctly visible. Using the layer scale dependency, we can make these objects invisible at this scale.

It is also possible to alter visibility with scale at the geographic object level within a layer. For example, you may wish to show only the major roads at a small scale. However, this will generally require more effort to produce. Object-level visibility can be driven by attributes already existing or created for the purpose of scale dependency. It can also be defined according to the geometric characteristics of an object, such as its area. In general, smaller features should not be viewable at lower scales.

A common way to achieve layer dependency at the object level using the whole-layer dependency is to select objects that match the given criteria and create new layers from these. Scale dependency can be applied to the subsets of the object class now contained in this separate layer.

You will want to set the layer scale dependency in accordance with scale ratios that conform to those that are commonly used. These are based on some assumptions, including those about the resolution of the tiled image (96 dpi) and the size of the tile (256px x 265px).

Zoom	Object extent	Scale at 96 dpi
0	Entire planet	1 : 59165752759.16
1		1 : 295,829,355.45
2		1 : 147,914,677.73
3		1 : 73,957,338.86
4		1 : 36,978,669.43
5	Country	1 : 18,489,334.72
6		1 : 9,244,667.36
7		1 : 4,622,333.68
8	State	1 : 2,311,166.84
9		1 : 1,155,583.42
10	Metropolitan	1 : 577,791.71
11		1 : 288,895.85
12	City	1 : 144,447.93
13		1 : 72,223.96
14	Town	1 : 36,111.98
15		1 : 18,055.99
16	Minor road	1 : 9,028.00
17		1 : 4,514.00
18	Sidewalks	1 : 2,257.00

The label conflict

Labels are commonly separated from the basemap layer itself. One reason for this is that if labels are included in the basemap layer, they will be obscured by the operational layer displayed above it. Another reason is that tile caching sometimes does not properly handle labels, causing fragments of labels to be left missing. Labels should also be displayed with their own scale dependency, filtering out only the most important labels at smaller scales. If you have many layers and objects to be labeled, this may be a good use case for a map server or at least a rendering engine such as Mapnik.

The polygon label conflict resolution

To achieve conflict resolution between label layers on our map output, we will convert the geographic objects to be labeled to centroids—points in the middle of each object—which will then be displayed along with the label field as a label layer.

1. Convert objects to points through the following steps:

    1. Navigate to **Vector | Geometry Tools | Polygon Centroids**.

    2. If the polygons are in a database, create an SQL view where the polygons are stored, as shown in the following code:

    ```
    CREATE VIEW AS
    SELECT polygon_class.label, st_centroid
      (polygon_class.geography) AS geography
      FROM polygon_class;
    ```

2. Create a layer corresponding to the labels in the map server or renderer.

3. Add any adjustments via the SLD or whichever style markup you will use. The GeoServer implementation is particularly good at resolving conflicts and improving placement.

Chapter 7, Mapping for Enterprises and Communities, includes a more detailed blueprint for creating a labeling layer with a cartographically enhanced placement and conflict resolution using SLD in GeoServer.

The characteristics of the basemap will affect the range of interaction, panning, and zooming in the map interface. You will want a basemap that covers the extent of the area to be seen on the map interface and probably restrict the interface to a region of interest. This way, someone viewing a collection of buildings in a corner of one state does not get lost panning to the opposite corner of another state! When you cache your basemap, you will want to indicate that you wish to cache to this extent. Similarly, viewable scales will be configured at the time your basemap is cached, and you'll want to indicate which these are. This affects the incremental steps, in which the zoom tool increases or decreases the map scale.

Tile caches

The best way to cache your basemap data so that it quickly loads is to save it as individual images. Rather than requiring a potentially complicated rendering by the browser of many geometric features, a few images corresponding to the scale and extent to which they are viewed can be quickly transferred from client to server and displayed. These prerendered images are referred to as tiles because these square images will be displayed seamlessly when the basemap is requested. This is now the standard method used to prepare data for web mapping. In this book, we will cover two tools to create tile caches: QTiles plugin (*Chapter 1, Exploring Places – from Concept to Interface*) and TileMill/MBTiles (*Chapter 7, Mapping for Enterprises and Communities*).

	Configuration time	Execution time	Visual quality	Stored in a single file	Stored as image directories	Suitable for labels
QTiles Plugin	1	3	3	No	Yes	No
GDAL2Tiles.py	2	1	2	No	Yes	No
TileMill/ MBTiles	3	2	1	Yes	No	Yes
GeoServer/ GWC	3	2	1	No	No	Yes

You will need to pay some attention to the scheme for tile storage that is used. The .mbtiles format that TileMill uses is a SQLite database that will need to be read with a map interface that supports it, such as Leaflet. The QTiles plugin and GDAL2Tiles.py use an *XYZ* tile scheme with hierarchical directories based on row (*X*), column (*Y*), and zoom (*Z*) respectively with the origin in the top-left corner of the map. This is the most popular tiling scheme. The TMS tiling scheme sometimes used by GeoServer open source map server (which supports multiple schemes/service specifications) and that accepted by OSGeo are almost identical; however, the origin is at the bottom-left of the map. This often leads to some confusing results. Note that zoom levels are standardized according to the tile scheme tile size and resolution (for example, 256 x 256 pixels)

Generating and testing a simple directory-based tile cache structure

We will now use the QTiles plugin to generate a directory-based *ZYX* tile scheme cache. Perform the following steps:

1. Install QTiles and the TileLayer plugin.
    -   QTiles is listed under the experimental plugins. You must alter the plugin settings to show experimental plugins. Navigate to **Plugins | Manage and Install Plugins | Settings | "Show also experimental plugins"**.

2. Run QTiles, creating a new `mytiles` tileset with a minimum zoom of `14` and maximum of `16`.

3. You'll realize the value of this directory in the next example.

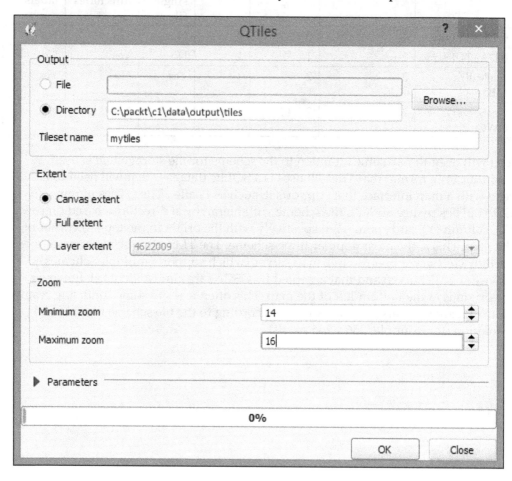

4. You can test and see whether the tiles were created by looking under the directory where you created them. They will be under the directory in the numbered subdirectories given by their *Z*, *X*, and *Y* grid positions in the tiling scheme. For example, here's a tile at 15/9489/12442.png. That's 15 zoom, 9489 longitude in the grid scheme, and 12442 latitude in the grid scheme.

You will now see the following output:

Create a layer description file for the TileLayer plugin

Create a layer description file with a `tsv` (tab delimited) extension in the UTF-8 encoding. This is a universal text encoding that is widely used on the Web and is sometimes needed for compatibility.

 Note that the last six parameters are optional and may prevent missing tiles. In the following example, I will use only the z parameters, `zmin` and `zmax`, related to map zoom level.

1. Add text in the following form, containing all tile parameters, to a new file:

   ```
   title credit url yOriginTop [zmin zmax xmin ymin xmax ymax]
   ```

   ○ For example, `mytiles  me  file:///c:/packt/c1/data/output/tiles/ mytiles/{z}/{x}/{y}.png  1  14  16`.

   ○ In the preceding example, the description file refers to a local Windows file system path, where the tiled `.png` images are stored.

2. Save `mytiles.tsv` to the following path:

```
[YOUR HOME DIRECTORY]/.qgis2///python/plugins/
    TileLayerPlugin/layers
```

- For me, on Windows, this was `C:\Users\[user]\.qgis2\python\` `plugins\TileLayerPlugin\layers`.

[Note that `.qgis2` may be a hidden directory on some systems. Make sure to show the hidden directories/files.]

- The path for the location to save your TSV file can be found or set under **Web | TileLayer Plugin | Add Tile Layer | Settings | External layer definition directory**

Preview it with the TileLayer plugin. You should be able to add the layer from the TilerLayerPlugin dialog. Now that the layer description file has been added to the correct location, let's go to **Web TileLayerPlugin | Add Tile Layer**:

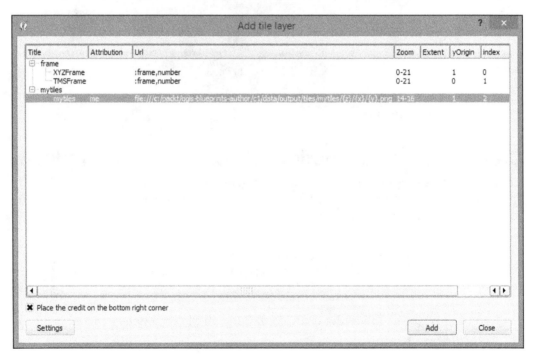

After selecting the layer, click on **Add**. Your tiles will look something like the following image:

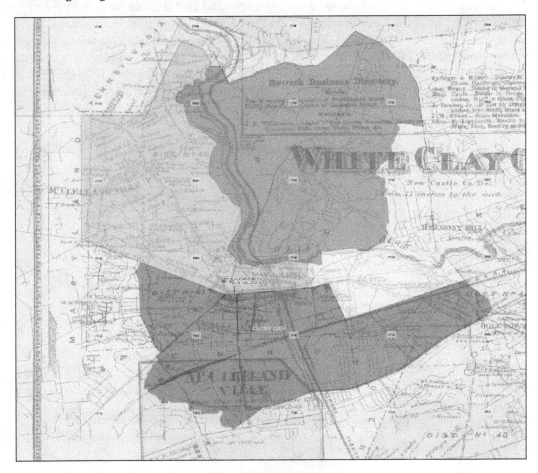

 Note the credit value in the lower-right corner of each tile.

Summary

In this chapter, you learned the necessary background and took steps to get up and running with QGIS. We performed ETL on the location-based data to geospatially integrate it with our GIS project. You learned the fundamental GIS visualization techniques around layer style and labeling. Finally, after some consideration around the nature of basemaps, we produced a tile cache that we could preview in QGIS. In the next chapter, we will use raster analysis to produce an operational layer for interaction within a simple web map application.

Identifying the Best Places

2

In this chapter, we will take a look at how the raster data can be analyzed, enhanced, and used for map production. Specifically, you will learn to produce a grid of the suitable locations based on the criteria values in other grids using raster analysis and map algebra. Then, using the grid, we will produce a simple click-based map. The end result will be a site suitability web application with click-based discovery capabilities. We'll be looking at the suitability for the farmland preservation selection.

In this chapter, we will cover the following topics:

- Vector data ETL for raster analysis
- Batch processing
- Raster analysis concepts
- Map algebra
- Additive modeling
- Proximity analysis
- Raster data ETL for vector publication
- Leaflet map application publication with qgis2leaf

Vector data – Extract, Transform, and Load

Our suitability analysis uses map algebra and criteria grids to give us a single value for the suitability for some activity in every place. This requires that the data be expressed in the raster (grid) format. So, let's perform the other necessary ETL steps and then convert our vector data to raster.

We will perform the following actions:

- Ensure that our data has identical spatial reference systems. For example, we may be using a layer of the roads maintained by the state department of transportation and a layer of land use maintained by the department of natural resources. These layers must have identical spatial reference systems or be transformed to have identical systems.

- Extract geographic objects according to their classes as defined in some attribute table field if we want to operate on them while they're still in the vector form.

- If no further analysis is necessary, convert to raster.

Loading data and establishing the CRS conformity

It is important for the layers in this project to be transformed or projected into the same geographic or projected coordinate system. This is necessary for an accurate analysis and for publication to the web formats. Perform the following steps for this:

1. Disable 'on the fly' projection if it is turned on. Otherwise, 'on the fly' will automatically project your data again to display it with the layers that are already in the Canvas.

    1. Navigate to **Settings | Options** and configure the settings shown in the following screenshot:

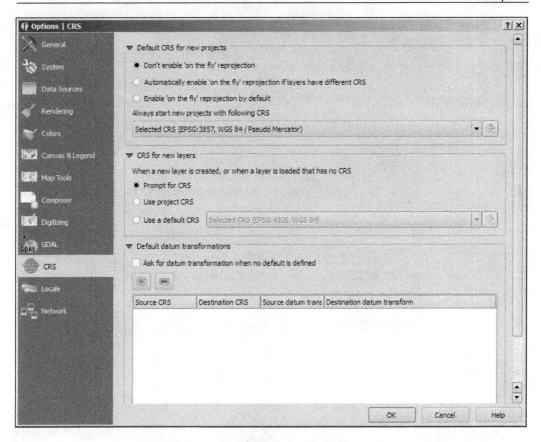

2. Add the project layers:

    2. Navigate to **Layer** | **Add Layer** | **Vector Layer**.

    3. Add the following layers from within `c2/data/original`.

        Applicants

        County

        Easements

        Land use

        Roads

You can select multiple layers to add by pressing *Shift* and clicking on the contiguous files or pressing *Ctrl* and clicking on the noncontiguous files.

3. Import the Digital Elevation Model from `c2/data/original/dem/dem.tif`.

    1. Navigate to **Layer** | **Add Layer** | **Raster Layer**.

    2. From the `dem` directory, select `dem.tif` and then click on **Open**.

4. Even though the layers are in a different CRS, QGIS does not warn us in this case. You must discover the issue by checking each layer individually. Check the CRS of the county layer and one other layer:

    1. Highlight the county layer in the **Layers** panel.

    2. Navigate to **Layer** | **Properties**.

    3. The CRS is displayed under the **General** tab in the **Coordinate reference system** section:

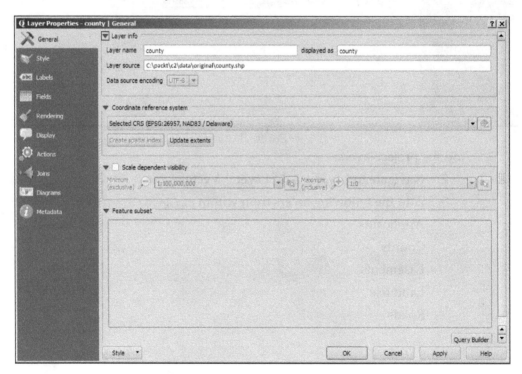

 Note that the county layer is in **EPSG: 26957**, while the others are in **EPSG: 2776**.

5. Follow the steps in *Chapter 1, Exploring Places – from Concept to Interface*, for transformation and projection. We will transform the county layer from EPSG:26957 to EPSG:2776.

    1. Navigate to **Layer** | **Save as** | **Select CRS**.

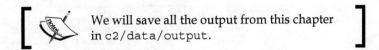

> We will save all the output from this chapter in c2/data/output.

To prepare the layers for conversion to raster, we will add a new generic column to all the layers populated with the number 1. This will be translated to a Boolean type raster, where the presence of the object that the raster represents (for example, roads) is indicated by a cell of 1 and all others with a zero. Follow these steps for the applicants, easements, and roads:

1. Navigate to **Layer** | **Toggle Editing**.
2. Then, navigate to **Layer** | **Open Attribute Table**.
3. Add a column with the button at the top of the **Attribute table** dialog.
4. Use value as the name for the new column and the following data format options:

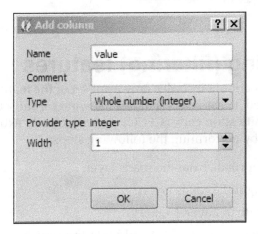

5. Select the new column from the dropdown in the **Attribute table** and enter 1 into the value box:

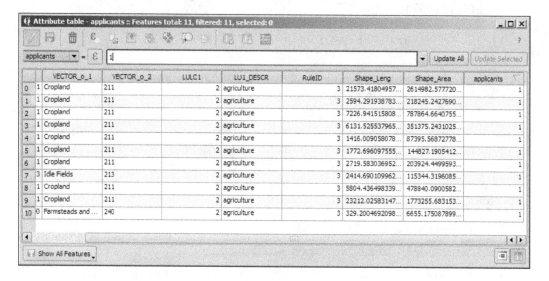

6. Click on **Update All**.
7. Navigate to **Layer | Toggle Editing**.
8. Finally, save.

The extracting (filtering) features

Let's suppose that our criteria includes only a subset of the features in our roads layer – major unlimited access roads (but not freeways), a subset of the features as determined by a **classification code** (CFCC). To temporarily extract this subset, we will do a layer query by performing the following steps:

1. Filter the major roads from the roads layer.
   1. Highlight the roads layer.
   2. Navigate to **Layer | Query**.
   3. Double-click on **CFCC** to add it to the expression.
   4. Click on the = operator to add to the expression
   5. Under the **Values** section, click on **All** to view all the unique values in the **CFCC** field.
   6. Double-click on **A21** to add this to the expression.
   7. Do this for all the codes less than A36. Include A63 for highway on-ramps.

8. Your selection code will look similar to this:

```
"CFCC" = 'A21' OR "CFCC" = 'A25' OR "CFCC" =
    'A31' OR "CFCC" = 'A35' OR "CFCC" = 'A63'
```

9. Click on **OK**, as shown in the following screenshot:

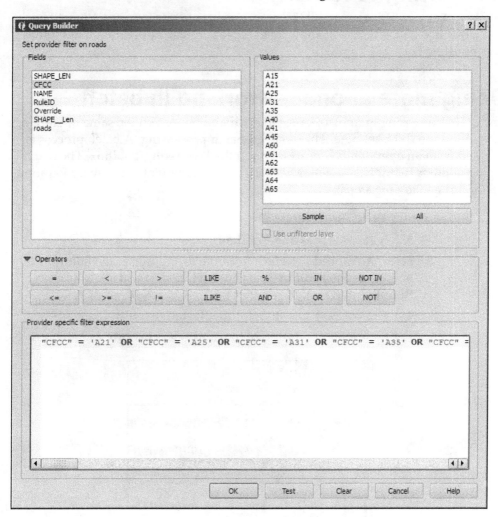

2. Save the roads layer as a new layer with only the selected features (`major_roads`) in `c2/data/output`.

To clear a layer filter, return to the query dialog on the applied layer (highlight it in the **Layers** pane; navigate to **Layer** | **Query** and click on **Clear**).

3. Repeat these steps for the `developed` (`LULC1` = 1) and `agriculture` (`LULC1` = 2) land uses (separately) from the `landuse` layer.

Converting to raster

In this section, we will convert all the needed vector layers to raster. We will be doing this in batch, which will allow us to repeat the same operation many times over multiple layers.

Doing more at once—working in batch

The QGIS Processing Framework provides capabilities to run the same operation many times on different data. This is called **batch processing**. A batch process is invoked from an operation's context menu in the **Processing Toolbox**. The batch dialog requires that the parameters for each layer be populated for every iteration. Perform the following steps:

1. Convert the vector layers to raster.

    1. Navigate to **Processing Toolbox**.

    2. Select **Advanced Interface** from the dropdown at the bottom of **Processing Toolbox** (if it is not selected, it will show as **Simple Interface**).

    3. Type `rasterize` to search for the **Rasterize** tool.

    4. Right-click on the **Rasterize** tool and select **Execute as batch process**:

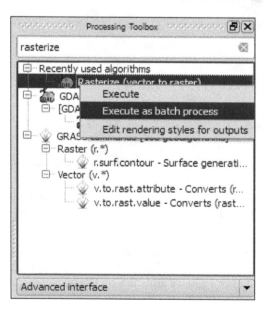

5. Fill in the **Batch Processing** dialog, making sure to specify the parameters as follows:

Parameter	Value
Input layer	(For example, `roads`)
Attribute field	`value`
Output raster size	**Output resolution in map units per pixel**
Horizontal	30
Vertical	30
Raster type	**Int16**
Output layer	(For example, `roads`)

The following images show how this will look in QGIS:

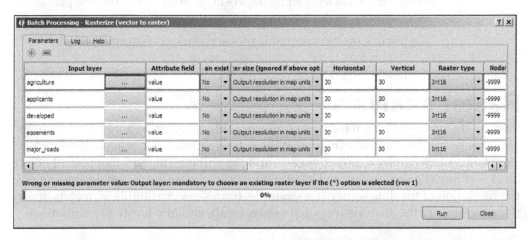

6. Scroll to the right to complete the entry of parameter values.

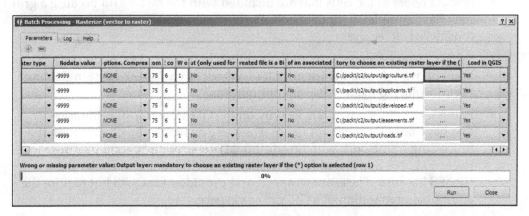

2. Organize the new layers (optional step).

   ○ Batch sometimes gives unfriendly names based on some bug in the dialog box.

   ○ Change the layer names by doing the following for each layer created by batch:

   1. Highlight the layer.

   2. Navigate to **Layer | Properties**.

   3. Change the layer name to the name of the vector layer from which this was created (for example, `applicants`). You should be able to find a hint for this value in the layer properties in the layer source (name of the `.tif` file).

   4. Group the layers:

      Press *Shift* + click on all the layers created by batch and the previous `roads` raster, in the **Layers** panel.

      Right-click on the selected layers and click on **Group selected**.

Raster analysis

Raster data, by organizing the data in uniform grids, is useful to analyze continuous phenomena or find some information at the subobject level. We will use continuous elevation and proximity data in this case, and we will look at the subapplicant object level — at the 30 meter-square cell level. You would choose a cell size depending on the resolution of the data source (for example, from sensors roughly 30 meters apart), the roughness of the analysis (regional versus local), and any hardware limitations.

First, let's make a few notes about raster data:

- Nodata refers to the cells that are included with the raster grid because a grid can't have completely undefined cells; however, these cells should really be considered *off the layer*.

- QGIS's raster renderer is more limited than in its proprietary competitors. You will want to use the **Identify** tool as well as custom styles (**Singleband Pseudocolor**) to make sense of your outputs.

- In this example, we will rely heavily on the GDAL and SAGA libraries that have been wrapped for QGIS. These are available directly through the processing framework with no additional preparation beyond the ordinary raster ETL. For additional functionality, you will want to consider the GRASS libraries. These are wrapped and provided for QGIS but require the additional preparation of a GRASS workspace.

Now that all our data is in the raster format, we can work through how to derive information from these layers and combine this information in order to select the best sites.

Map algebra

Map algebra is a useful concept to work with multiple raster layers and analysis steps, providing arithmetic operations between cells in aligned grids. These produce an output grid with the respective value of the arithmetic solution for each set of cells. We will be using map algebra in this example for additive modeling.

Additive modeling

Now that all our data is in the raster format, we can begin to model for the purpose of site selection. We want to discover which cells are best according to a set of criteria which has either been established for the domain area (for example, the agricultural conservation site selection) by convention or selected at the time of modeling. Additive modeling refers to this process of adding up all the criteria and associated weights to find the best areas, which will have the greatest value.

In this case, we have selected some criteria that are loosely known to affect the agricultural conservation site selection, as shown in the following table:

Layer	Criteria	Rule
applicants	Is applicant	
easements	Proximity	< 2000 m
landuse (agriculture)	Land use, proximity	< 100 m
dem	Slope	=> 2 and <= 5, average
landuse (developed)	Land use, proximity	> 500 m
roads	Proximity	> 100m

Proximity

The **Proximity** grid tool will generate a layer of cells with each cell having a value equal to its distance from the nearest non-nodata cell in another grid. The distance value is given in the CRS units of the other grid. It also generates direction and allocation grids with the direction and ID of the nearest nodata cell.

Creating a proximity to the easements grid

Perform the following steps:

1. Navigate to **Processing Toolbox**.

2. Search for `proximity` in this toolbox. Ensure that you have the **Advanced Interface** selected.

3. Once you've located the **Proximity** grid tool under **SAGA**, double-click on it to run it.

4. Select **easements** for the **Features** field.

5. Specify an output file for **Distance** at `c2/data/output/easements_prox.tif`.

6. Uncheck **Open output file after running algorithm** for the other two outputs, as shown in the following screenshot:

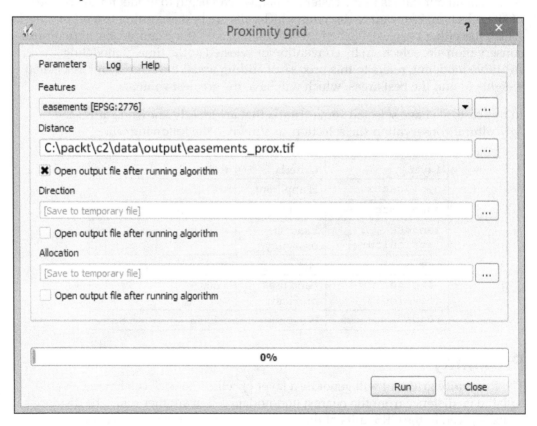

The resulting grid is of the distance to the closest easement cell.

7. Repeat these steps to create proximity grids for `agriculture`, `developed`, and `roads`. Finally, you will see the following output:

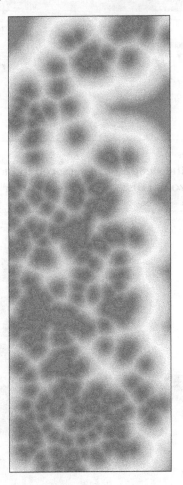

Slope

The **Slope** command creates a grid where the value of each cell is equal to the upgradient slope in percent terms. In other words, it is equal to how steep the terrain is at the current cell in the percentage of rise in elevation unit per horizontal distance unit. Perform the following steps:

1. Install and activate the Raster Terrain Analysis plugin if you have not already done so.

2. Navigate to **Raster | Terrain Analysis | Slope**.

3. Select **dem**, the Digital Elevation Model, for the **Elevation layer** field.

4. Save your output in c2/data/output. You can keep the other inputs as default.

5. The output will be the steepness of each cell in the percentage of of vertical elevation over horizontal distance ("rise over run").

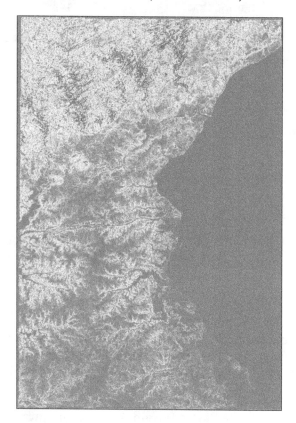

Combining the criteria with Map Calculator

1. Ensure that all the criteria grids (proximity, agriculture, developed, road, and slope) appear in the **Layers** panel. If they don't, add them.

2. Bring up the **Raster calculator** dialog.

    1. Navigate to **Raster | Raster calculator**

3. Enter the map algebra expression.

    ° Add the raster layers by double-clicking on them in the **Raster bands** selection area

    ° Add the operators by typing them out or clicking on the buttons in the operators area

    ° The expression entered should be as follows:

```
("slope@1" < 8) + ("applicants@1" = 1) +
  ("easement_prox@1"<2000) + ("roads_prox@1">100) +
  ("developed_prox@1" > 500) + ("agriculture@1" < 100)
```

 @1 refers to the first and only band of the raster.

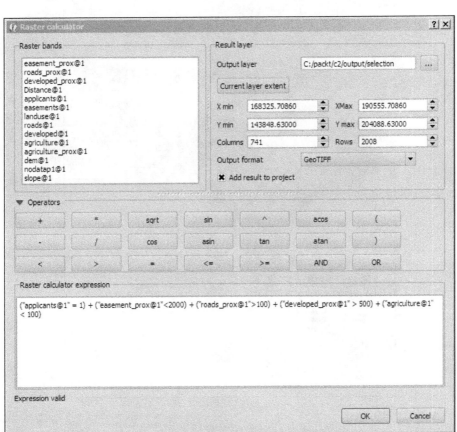

4. Add a name and path for the output file and hit *Enter*.

5. You may need to set a style if it seems like nothing happened. By default, the nonzero value is set to display in white (the same color as our background).

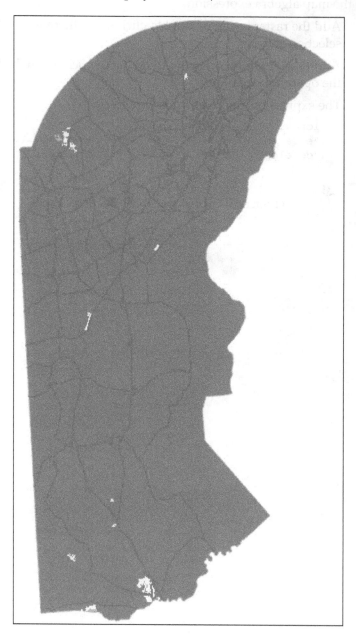

Here's a close up of the preceding map image so that you can see the variability in suitability:

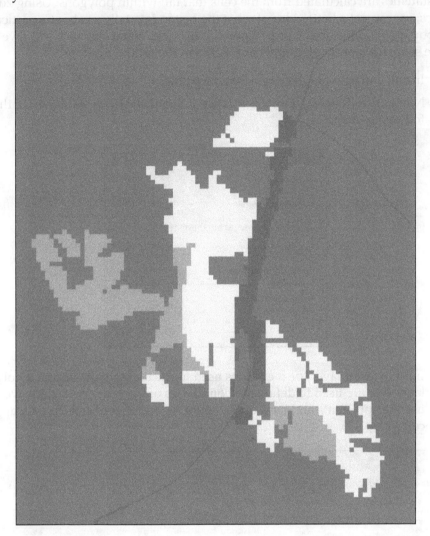

In the preceding screenshot, cells are scored as follows:

- Green = 5 (high)
- Yellow = 4 (middle)
- Red = 3 (low)

Zonal statistics

Zonal statistics are calculated from the cells that fall within polygons. Using zonal statistics, we can get a better idea of what the raster data tells us about a particular cell group, geographic object, or polygon. In this case, zonal statistics will give us an average score for a particular applicant. Perform the following steps:

1. Install and activate the Zonal Statistics plugin.

2. Navigate to **Raster | Zonal Statistics | Zonal statistics**, as shown in the following image:

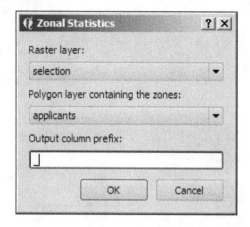

3. Input a raster layer for the values used to calculate a statistic and a polygon layer that are used to define the boundaries of the cells used. Here, we will use the applicants and land use to count the number of cells in each applicant cell group.

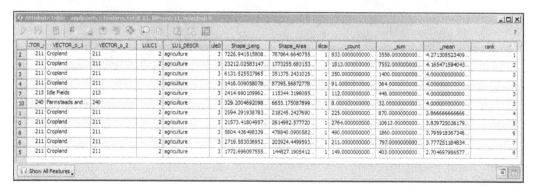

	:TOR_	VECTOR_o_1	VECTOR_o_2	LULC1	LU1_DESCR	ulell	Shape_Leng	Shape_Area	slica	_count	_sum	_mean	rank
2	211	Cropland	211	2	agriculture	3	7226.941515808...	787864.6640755...	1	833.0000000000...	3558.000000000...	4.271308523409...	1
9	211	Cropland	211	2	agriculture	3	23212.02583147...	1773255.683153...	1	1813.000000000...	7552.000000000...	4.165471594043...	2
3	211	Cropland	211	2	agriculture	3	6131.525537965...	351375.2431025...	1	350.0000000000...	1400.000000000...	4.000000000000...	3
4	211	Cropland	211	2	agriculture	3	1416.009058078...	87395.56872778...	1	91.00000000000...	364.0000000000...	4.000000000000...	3
7	213	Idle Fields	213	2	agriculture	3	2414.690109962...	115344.3196085...	1	112.0000000000...	448.0000000000...	4.000000000000...	3
10	240	Farmsteads and ...	240	2	agriculture	3	329.2004692098...	6655.175087899...	1	8.000000000000...	32.00000000000...	4.000000000000...	3
1	211	Cropland	211	2	agriculture	3	2594.291938783...	218245.2427690...	1	225.0000000000...	870.0000000000...	3.866666666666...	4
0	211	Cropland	211	2	agriculture	3	21573.41804957...	2614982.577720...	1	2764.000000000...	10613.00000000...	3.839725036179...	5
8	211	Cropland	211	2	agriculture	3	5804.436498339...	478840.0900582...	1	490.0000000000...	1860.000000000...	3.795918367346...	6
6	211	Cropland	211	2	agriculture	3	2719.583036952...	203924.4499593...	1	211.0000000000...	797.0000000000...	3.777251184834...	7
5	211	Cropland	211	2	agriculture	3	1772.696097555...	144827.1905412...	1	149.0000000000...	403.0000000000...	2.704697986577...	8

Show All Features

4. Create a `rank` field, editing each value manually according to the _mean field created by the zonal statistics step. This is a measure of the mean suitability per cell. We will use this field for a label to communicate the relative suitability to a general audience; so, we want a rank instead of the rough mean value.

5. Now, label the layer.

   1. Under **Layer Properties**, activate the **Labels** tab.

   2. Choose the `rank` field as the field to label.

   3. Add any other formatting, such as label placement and buffer (halo) using the inner tabs within the label tab dialog, as shown in the following screenshot:

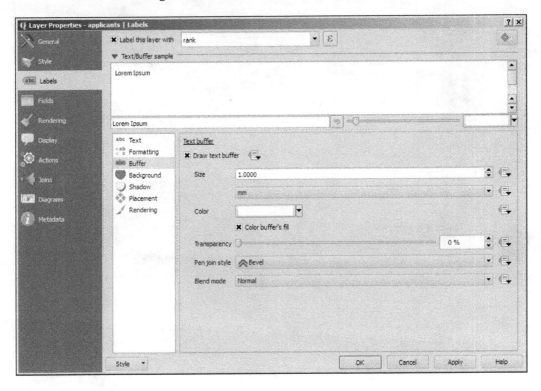

6. Add a style to the layer.

   1. Select the **Graduated** style.

   2. Select a suitable color ramp, number of classes, and classification type.

   3. Click on the **Classify** button, as shown in the following screenshot:

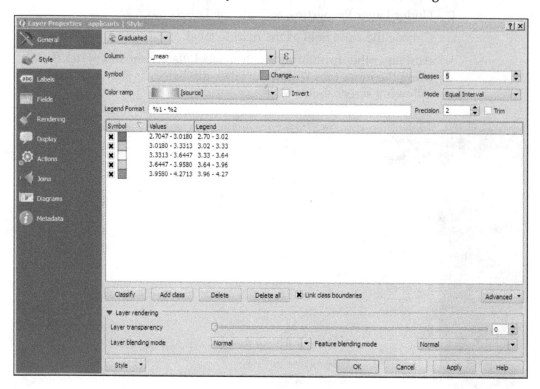

After you've completed these steps, your map will look something similar to this:

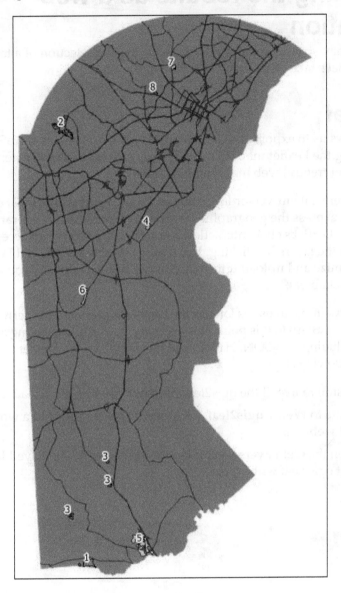

Publishing the results as a web application

Now that we have completed our modeling for the site selection of a farmland for conservation, let's take steps to publish this for the Web.

qgis2leaf

qgis2leaf allows us to export our QGIS map to web map formats (JavaScript, HTML, and CSS) using the Leaflet map API. Leaflet is a very lightweight, extensible, and responsive (and trendy) web mapping interface.

qgis2leaf converts all our vector layers to GeoJSON, which is the most common textual way to express the geographic JavaScript objects. As our operational layer is in GeoJSON, Leaflet's click interaction is supported, and we can access the information in the layers by clicking. It is a fully editable HTML and JavaScript file. You can customize and upload it to an accessible web location, as you'll understand in subsequent chapters.

qgis2leaf is very simple to use as long as the layers are prepared properly (for example, with respect to CRS) up to this point. It is also very powerful in creating a good starting application including GeoJSON, HTML, and JavaScript for our Leaflet web map. Perform the following steps:

1. Make sure to install the qgis2leaf plugin if you haven't already.

2. Navigate to **Web | qgis2leaf | Exports a QGIS Project to a working Leaflet webmap**.

3. Click on the **Get Layers** button to add the currently displayed layers to the set that qgis2leaf will export.

4. Choose a basemap and enter the additional details if so desired.

5. Select **Encode to JSON**.

These steps will produce a map application similar to the following one. We'll take a look at how to restore the labels in the next chapter:

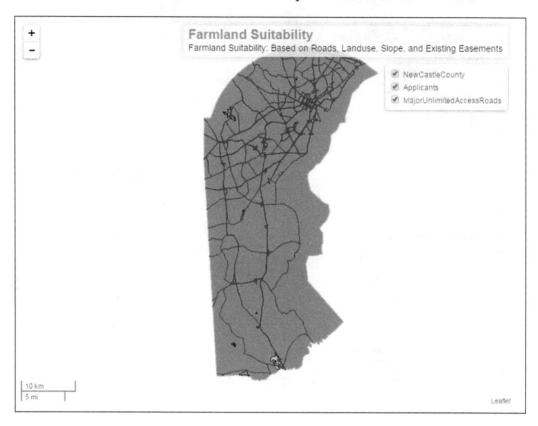

Summary

In this chapter, using the site selection example, we covered basic vector data ETL, raster analysis, and web map creation. We started with vector data, and after unifying CRS, we prepared the attribute tables. We then filtered and converted it to raster grids using batch processing. We also considered some fundamental raster concepts as we applied proximity and terrain analysis. Through map algebra, we combined these results for additive modeling site selection. We prepared these results, which required conversion to vector, styling, and labeling. Finally, we published the prepared vector output with qgis2leaf as a simple Leaflet web map application with a strong foundation for extension. In the next chapter, you will learn more about raster analysis and web application publishing with a hydrological modeling example.

3

Discovering Physical Relationships

In this chapter, we will create an application for a raster physical modeling example. First, we'll use a raster analysis to model the physical conditions for some basic hydrological analysis. Next, we'll redo these steps using a model automation tool. Then, we will attach the raster values to the vector objects for an efficient lookup in a web application. Finally, we will use a cloud platform to enable a dynamic query from the client-side application code. We will take a look at an environmental planning case, providing capabilities for stakeholders to discover the upstream toxic sites.

In this chapter, we will cover the following topics:

- Hydrological modeling
- Workflow automation with graphical models
- Spatial relationships for a performant access to information
- The NNJoin plugin
- The CartoDB cloud platform
- Leaflet SQLQueries using an external API:CartoDB

Hydrological modeling

The behavior of water is closely tied with the characteristics of the terrain's surface — particularly the values connected to elevation. In this section, we will use a basic hydrological model to analyze the location and direction of the hydrological network — streams, creeks, and rivers. To do this, we will use a digital elevation model and a raster grid, in which the value of each cell is equal to the elevation at that location. A more complex model would employ additional physical parameters (e.g., infrastructure, vegetation, etc.). These modeling steps will lay the necessary foundation for our web application, which will display the upstream toxic sites (brownfields), both active and historical, for a given location.

There are a number of different plugins and Processing Framework algorithms (operations) that enable hydrological modeling. For this exercise, we will use SAGA algorithms, of which many are available, with some help from GDAL for the raster preparation. Note that you may need to wait much longer than you are accustomed to for some of the hydrological modeling operations to finish (approximately an hour).

Preparing the data

Some work is needed to prepare the DEM data for hydrological modeling. The DEM path is `c3/data/original/dem/dem.tif`. Add this layer to the map (navigate to **Layer** | **Add Layer** | **Add Raster Layer**). Also, add the county shapefile at `c3/data/original/county.shp` (navigate to **Layer** | **Add Layer** | **Add Vector Layer**).

Filling the grid sinks

Filling the grid sinks smooths out the elevation surface to exclude the unusual low points in the surface that would cause the modeled streams to — unrealistically — drain to these local lows instead of to larger outlets. The steps to fill the grid sinks are as follows:

1. Navigate to **Processing Toolbox** (**Advanced Interface**).
2. Search for **Fill Sinks** (under **SAGA** | **Terrain Analysis** | **Hydrology**).
3. Run the **Fill Sinks** tool.
4. In addition to the default parameters, define **DEM** as `dem` and **Filled DEM** as `c3/data/output/fill.tif`.

5. Click on **Run**, as shown in the following screenshot:

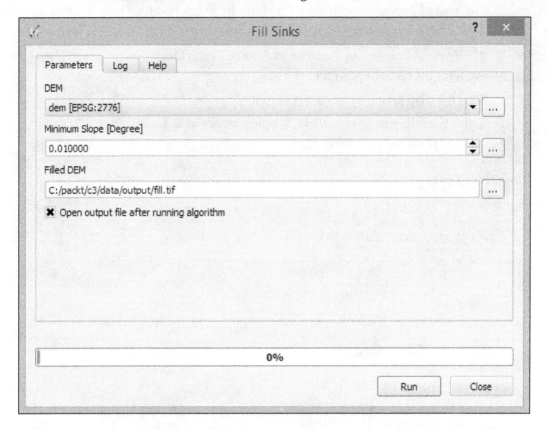

Clipping the grid to study the area by mask layer

By limiting the raster processing extent, we exclude the unnecessary data, improving the speed of the operation. At the same time, we also output a more useful grid that conforms to our extent of interest. In QGIS/SAGA, in order to limit the processing to a fixed extent or area, it is necessary to eliminate those cells from the grid – in other words, the setting cells outside the area or extent, which are sometimes referred to as **NoData** (or no-data, and so on) in raster software, to a null value.

> Unlike ArcGIS or GRASS, the SAGA package under QGIS does not have any capability to set an extent or area within which we want to limit the raster processing.

In QGIS, the raster processing's extent limitation can be accomplished using a vector polygon or a set of polygons with the **Clip raster by mask layer** tool. By following the given steps, we can achieve this:

1. Navigate to **Processing Toolbox** (**Advanced Interface**).

2. Search for **Mask** (under **GDAL | Extraction**).

3. Run the **Clip raster by mask layer** tool.

4. Enter the following parameters, keeping others as default:

    ° **Input layer**: This is the layer corresponding to `fill.tif`, created in the previous **Fill Sinks** section

    ° **Mask layer**: `county`

    ° **Output layer**: `c3/data/output/clip.tif`

5. Click on **Run**, as shown in the following screenshot:

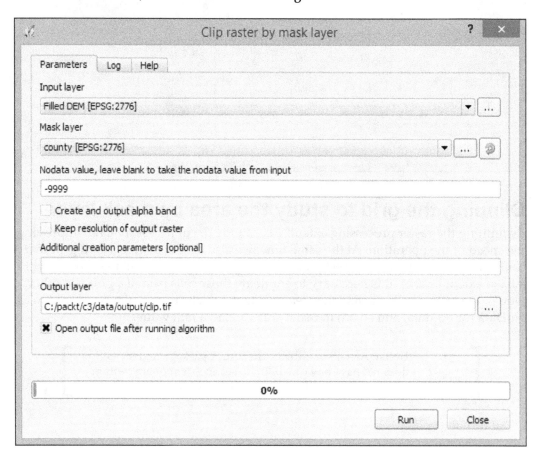

This function is not available in some versions of QGIS for Mac OS.

The output from **Clip by mask layer** tool, showing the grid clipped to the county polygon, will look similar to the following image (the black and white color gradient or mapping to a null value may be reversed):

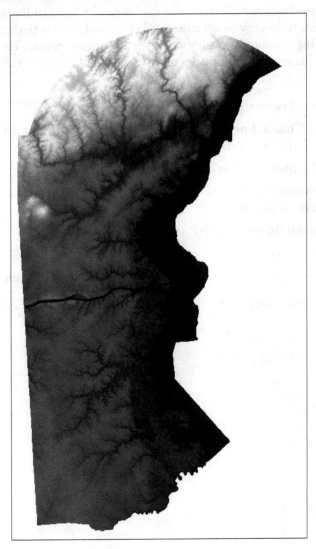

Modeling the hydrological network based on elevation

Now that our elevation grid has been prepared, it is time to actually model the hydrological network location and direction. To do this, we will use **Channel network and drainage basins**, which only requires a single input: the (filled and clipped) elevation model. This tool will produce the hydrological lines using a Strahler Order threshold, which relates to the hierarchy level of the returned streams (for example, to exclude very small ditches) The default of 5 is perfect for our purposes, including enough hydrological lines but not too many. The results look pretty realistic. This tool also produces many additional related grids, which we do not need for this project. Perform the following steps:

1. Navigate to **Processing Toolbox (Advanced Interface)**.
2. Search for **Channel network and drainage basins** (under **SAGA | Terrain Analysis | Hydrology**).
3. Run the **Channel network and drainage basins** tool.
4. In the **Elevation** field, input the filled and clipped DEM, given as the output in the previous section.
5. In the **Threshold** field, keep it at the default value (5.0).
6. In the **Channels** field, input c3/data/output/channels.shp.

   Ensure that **Open output file after running algorithm** is selected

7. Unselect **Open output file after running algorithm** for all other outputs.

8. Click on **Run**, as shown in the following screenshot:

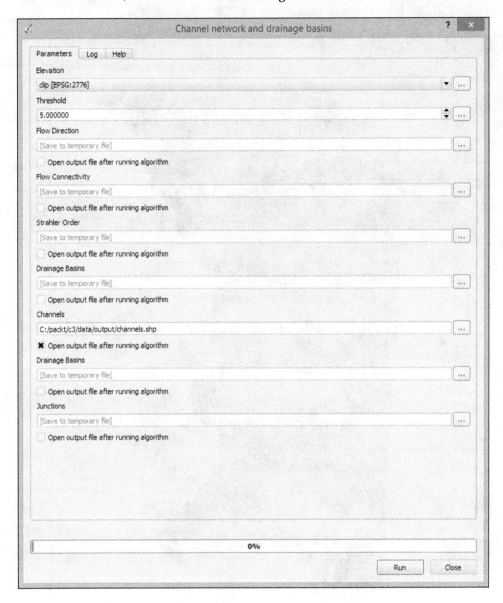

The output from the **Channel network and drainage basins**, showing the hydrological line location, will look similar to the following image:

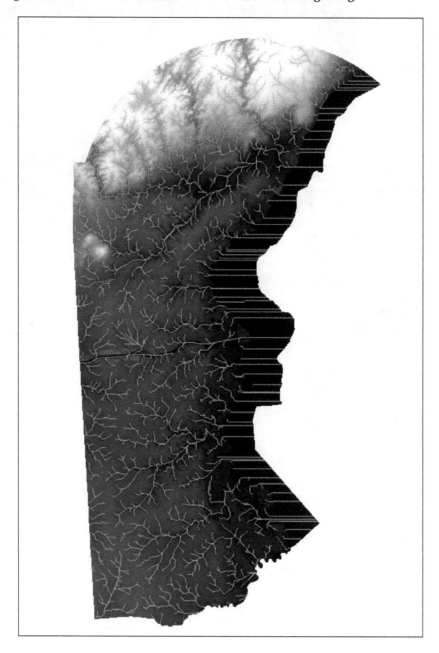

Workflow automation with the graphical models

Graphical Modeler is a tool within QGIS that is useful for modeling and automating workflows. It differs from batch processing in that you can tie together many separate operations in a processing sequence. It is considered a part of the processing framework. Graphical Modeler is particularly useful for workflows containing many steps to be repeated.

By building a graphical model, we can operationalize our hydrological modeling process. This provides a few benefits, as follows:

- Our modeling process is graphically documented and preserved
- The model can be rerun in its entirety with little to no interaction
- The model can be redistributed
- The model is parameterized so that we could rerun the same process on different data layers

Creating a graphical model

1. Bring up the **Graphical Modeler** dialog from the **Processing** menu.

    1. Navigate to **Processing | Graphical Modeler**.

2. Enter a model name and a group name.

3. Save your model under `c3/data/output/c3.model`.

    The dialog is modal and needs to be closed before you can return to other work in QGIS, so saving early will be useful.

Adding the input parameters

Some of the inputs to your model's algorithms will be the outputs of other model algorithms; for others, you will need to add a corresponding input parameter.

Adding the raster parameter – elevation

We will add the first data input parameter to the model so that it is available to the model algorithms. It is our original DEM elevation data. Perform the following steps:

1. Select the **Inputs** tab from the lower left corner of the **Processing modeler** display.

2. Drag **Raster layer** from the parameters list into the modeler pane.
    This parameter will represent our elevation grid (DEM).

3. Input `elevation` for **Parameter name**.

4. Click on **OK**, as shown in the following screenshot:

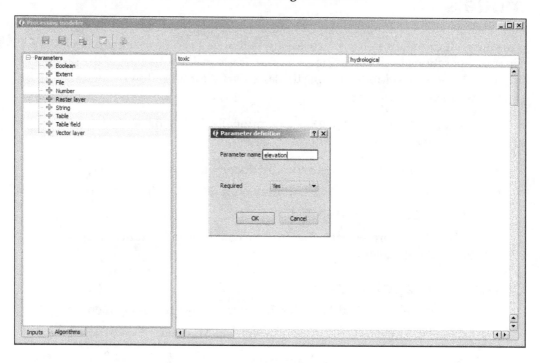

Adding the vector parameter – extent

We will add the next data input parameter to the model so that it is available to the model algorithms. It is our vector county data and the extent of our study.

1. Add a vector layer for our extent polygon (county). Make sure you select **Polygon** as the type, and call this parameter `extent`.

2. You will need to input a parameter name. It would be easiest to use the same layer/parameter names that we have been using so far, as shown in the following screenshot:

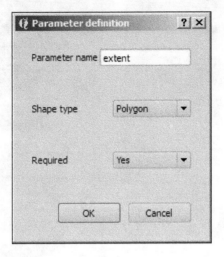

Adding the algorithms

The modeler connects the individual with their input data and their output data with the other algorithms. We will now add the algorithms.

Fill Sinks

The first algorithm we will add is **Fill Sinks**, which as we noted earlier, removes the problematic low elevations from the elevation data. Perform the following steps:

1. Select the **Algorithms** tab from the lower-left corner.

2. After you drag in an algorithm, you will be prompted to choose the parameters.

3. Use the search input to locate **Fill Sinks** and then open.

4. Select **elevation** for the DEM parameter and click on **OK**, as shown in the following screenshot:

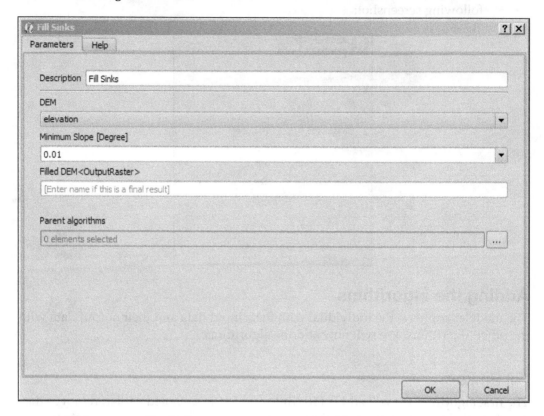

Clip raster

The next algorithm we will add is **Clip raster by mask layer**, which we've used to limit the processing extent of the subsequent raster processing. Perform the following steps:

1. Use the search input to locate **Clip raster by mask layer**.
2. Select **'Filled DEM' from algorithm 'Fill Sinks'** for the **Input layer** parameter.
3. Select **extent** for the **Mask layer** parameter.

4. Click on **OK**, accepting the other parameter defaults, as shown in the following screenshot:

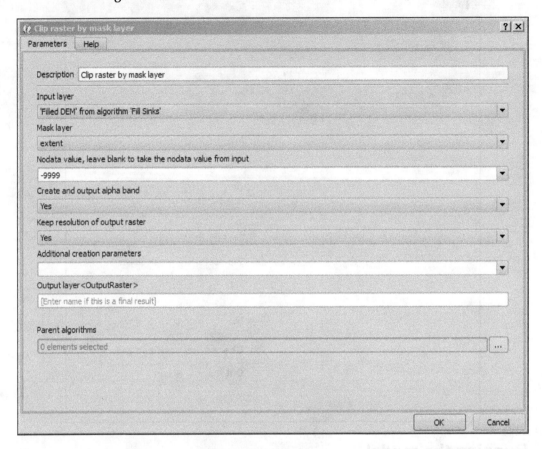

Channel network and drainage basins

The final algorithm we will add is **Channel network and drainage basins**, which produces a model of our hydrological network. Perform the following steps:

1. Use the search input to locate **Channel network and drainage basins**.

2. Select **'Output Layer' from algorithm 'Clip raster by mask layer'** for the **Elevation** parameter.

3. Click on **OK**, accepting the other parameter defaults.

4. Once you populate all the three algorithms, your model will look similar to the following image:

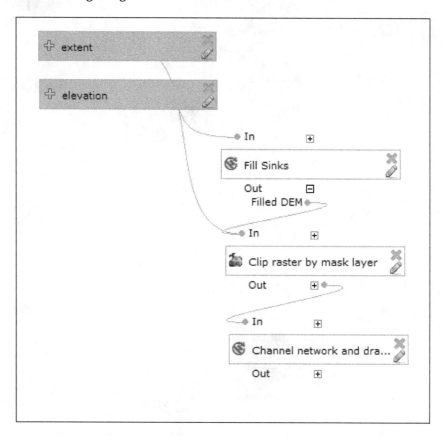

Running the model

Now that our model is complete, we can execute all the steps in an automated sequential fashion:

1. Run your model by clicking on the **Run model** button on the right-hand side of the row of buttons.

2. You'll be prompted to select values for the elevation and the extent input
    layer parameters you defined earlier. Select the **dem** and **county** layers for
    these inputs, respectively, as shown in the following screenshot:

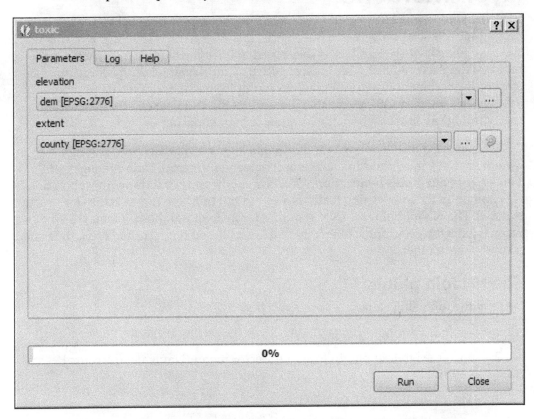

3. After you define and run your model, all the outputs you defined earlier
    will be produced. These will be located at the paths that you defined in the
    parameters dialog or in the model algorithms themselves.

 If you don't specify an output directory, the data will be saved to the
temp directory for the processing framework, for example:

`C:\Users\[YOURUSERNAME]\AppData\Local\Temp\processing\`

Now that we've completed the hydrological modeling, we'll look at a technique for
preparing our outputs for dynamic web interaction.

Spatial join for a performant operational layer interaction

A spatial join permanently relates two layers of geographic objects based on some geographic relationship between the objects. It is wise to do a spatial join in this way, and save to disk when possible, as the spatial queries can significantly increase the time of a database request. This is especially true for the tables with a large number of records or when your request involves multiple spatial or aggregate functions. In this case, we are performing a spatial join so that the end user can do the queries of the hydrological data based on the location of their choosing.

QGIS has less extensive options for the spatial join criteria than ArcGIS. The default spatial join method in QGIS is accessible via **Vector | Data Management Tools | Join attributes by location**. However, at the time of writing, this operation was limited to the intersecting features and did not offer the functionality for nearby features. The NNJoin plugin—**NN** standing for **nearest neighbor**—achieves what we want; it joins the geographic objects in two layers based on the criteria that they are nearest to each other.

The NNJoin plugin

Perform the following steps:

1. Install the NNJoin plugin.

2. Open the NNJoin plugin from the **Vector** menu (Navigate to **Vector | NNJoin**).

3. Specify the following parameters:
    - **Input vector layer**: Select this as **toxic**—layer of toxic sites.
    - **Join vector layer**: Select this as **Channels**—hydrological lines.
    - **Output layer**: Select this as **toxic_channels**. This operation only supports the output to memory. You'll need to click on **Save as** after running it to save it to disk.

4. Click on **OK**, as shown in the following screenshot:

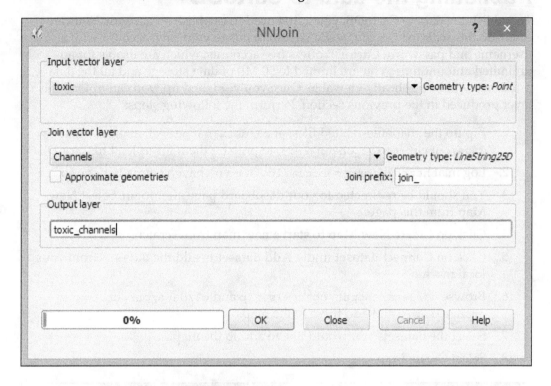

Now, in the **Layers** panel, right-click on the newly created **toxic_channels** layer and then on **Save as**. You should save this new file in the following path:

`c3/data/output/toxic_channels.shp`

The result of these steps will be a copy of the toxic layer with the columns from the nearest feature in the channels layer.

We've now completed all the data processing steps. It's now time to look at how we will host this data with an external cloud platform to enable the dynamic web query.

The CartoDB platform

CartoDB is a cloud-based GIS platform which provides data management, query, and visualization capabilities. CartoDB is based on Postgres/PostGIS in the backend, and one of the most exciting functions of this platform is the ability to pass spatial queries using PostGIS syntax via the URL and HTTP API.

Publishing the data to CartoDB

To publish the data to CartoDB, you'll first need to establish an account. You can easily do this with the Google Single sign-in or create your own account with a username and password. CartoDB offers free accounts, which are usable for an unlimited amount time. You are limited to 50 MB of data storage and all the data published will be publically viewable. Once you've signed up, you can upload the layer produced in the previous section. Perform the following steps:

1. Zip up the shapefile-related files, at c3/data/output/channels.* and c3/ data/output/toxic_channels.*, to prepare them for upload to CartoDB.

2. Log in at https://cartodb.com/login if you haven't already done so.

3. You should be redirected to your dashboard page after login. Select **New Map** from this page.

4. Click on **Create New Map** to start a new map from scratch.

5. Click on **Connect dataset** under **Add dataset** to add the datasets from your local machine.

6. Browse c3/data/output/channels.zip and c3/data/output/toxic_ channels.zip and add these to the map.

7. Select the datasets you would like to add to the map.

8. Select **Create Map**.

Preparing a CartoDB SQL Query

SQL is the lingua franca of database queries through which you can do anything from filtering to spatial operations to manipulating data on the database. There are slight differences in the way SQL works from one database system to the next one. The CartoDB SQL queries use valid Postgres/PostGIS syntax. For more information on Postgres/PostGIS SQL, check out the reference chapter in the manuals for Postgres (`http://www.postgresql.org/docs/9.4/interactive/reference.html`) for general functions and PostGIS (`http://postgis.net/docs/manual-2.1/reference.html`) for spatial functions.

There are a few different ways in which you can test your queries against CartoDB—each involving a different ease of input and producing a different result type.

Generating the test data

Our SQL query only requires one parameter that we do not know ahead of time: the coordinates of the user-selected click location. To simulate this interaction with QGIS and generate a coordinate pair, we will use the Coordinate Capture plugin. Perform the following steps:

1. Install the Coordinate Capture plugin if you have not already done so.
2. From the **Vector** menu, display the **Coordinate Capture** panel (navigate to **Vector | Coordinate Capture | Coordinate Capture**).
3. Select **Start capture** from the **Coordinate Capture** panel.

4. Click on a place on the map that you would expect to see upstream results for. In other words, based on the elevation surface, select a low point near a hydrological line; other hydrological objects should run down into that point, as shown in the following image:

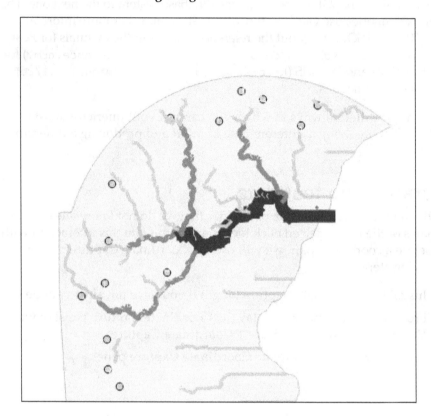

5. Record the coordinates displayed in the **Coordinate Capture** panel, as shown in the following screenshot:

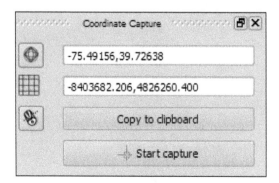

The CartoDB SQL view tab

Now, we will return to CartoDB in a web browser to run our first test.

While this method is probably the most straightforward in terms of data entry, it is limited to producing results via the map. There are no text results produced besides errors, which limits your ability to test and debug. Perform the following steps:

1. On your map, click on the tab corresponding to the **toxic_channels** layer. This is often accessed on the tab marked **2** on the right-hand side.

2. You should see the SQL view tab displayed by default with a SQL input area.

3. The SQL query given in the following section selects all the records from our joined table, which contains the location of toxic sites with their closest hydrological basin and stream order that fulfill the following criteria based on the coordinates we pass:

    ° It is in the same hydrological basin as the passed coordinates.

    ° It has a lower hydrological stream order than the closest stream to the passed coordinates.

    Recall that we generated test coordinates to pass with the **Coordinate Capture** plugin in the last step. Enter the following into the SQL area. This query will select all the fields from **toxic_channels** as expressed with the wildcard symbol (*) using various subqueries, joins, and spatial operations. The end result will show all the toxic sites that are upstream from the clicked point in its basin (code in c3/data/ original/query1.sql). Execute the following code:

```
SELECT toxic_channels.* FROM toxic_channels
INNER JOIN channels
ON toxic_channels.join_BASIN = channels.basin
WHERE toxic_channels.join_order <

(SELECT channels._order
FROM channels
WHERE
st_distance(the_geom, ST_GeomFromText
('POINT(-75.56111 39.72583)',4326))
IN (SELECT MIN(st_distance(the_geom,
ST_GeomFromText('POINT(-75.56111 39.72583)',4326)))
FROM channels x))

 AND toxic_channels.join_basin =

(SELECT channels.basin
FROM channels
```

```
WHERE
st_distance(the_geom, ST_GeomFromText
('POINT(-75.56111 39.72583)',4326))
IN (SELECT MIN(st_distance(the_geom,
ST_GeomFromText('POINT(-75.56111 39.72583)',4326)))
FROM channels x))
GROUP BY toxic_channels.cartodb_id
```

4. Select **Apply Query** to run the query.

If the query runs successfully, you should see an output similar to the following image:

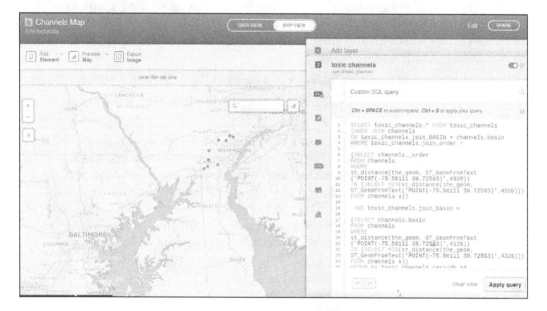

The following errors may confound the efforts to debug and test via the CartoDB SQL tab:

- **Error at the end of a statement**: A semi-colon, while valid, causes an error in this interface.
- **Does not contain cartodb_id**: The statement must explicitly contain a `cartodb_id` field so that it does not generate this error. However, this error does not typically affect the use through the API or URL parameters.
- **Does not contain the_geom**: The statement must explicitly contain a reference to the `the_geom` column even though this column is not visible within your `cartodb` table, to map the result.

Sometimes, the SQL input area is "sticky". If this happens, just "clear view".

The QGIS CartoDB plugin

Next, let's test the SQL from within QGIS using the QGIS CartoDB Plugin. Perform the following steps:

1. Install the QGIS CartoDB plugin, QGISCartoDB.

2. Open the SQL CartoDB dialog and navigate to **Web | CartoDB Plugin | Add SQL CartoDB Layer**.

3. Establish a connection to your CartoDB account:

    1. Click on **New**.

    2. Locate and enter your username and API key from your account in a browser. The query you ran earlier will be saved so you can do this in the open tab (if it is still open). Otherwise, navigate back to **CartoDB**. Your account name can be found in the URL when you are logged into CartoDB, where the username is in `username.cartodb.com/*`. You can find your API key by clicking on your avatar from your dashboard and selecting **Your API keys**.

    3. Click on **Save**, as shown in the following screenshot:

4. Now that you are connected to your CartoDB account, load tables from the CartoDB SQL Layer dialog.

5. Enter the preceding SQL statement in the SQL Query area. You can use the **Tables** section of the **Add CartoDB SQL Layer** dialog to view the field names and datatypes in your query.

6. Click on **Test Query** to test the syntax against CartoDB. Refer to the info box in the previous test section for some common confounding errors you may experience with the CartoDB SQL interface.

7. Click on **Add Layer** to add the result to QGIS.

The layer added from these steps will give you the location of the toxic sites upstream from the chosen coordinate. If you symbolized these locations with stars and streams according to their upstream/downstream rank, you would see something similar to the following image:

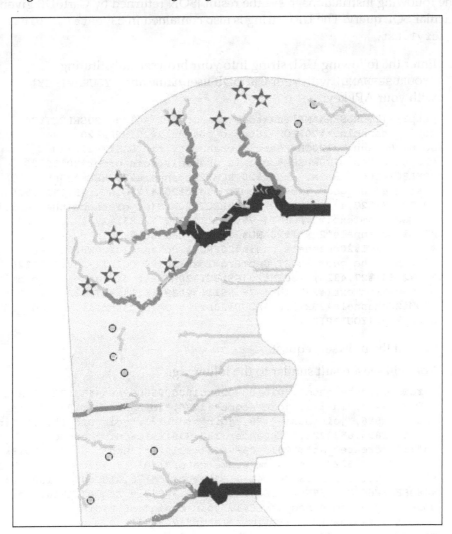

The CartoDB SQL API

If you want to see the actual contents returned by a CartoDB SQL query in the JSON format, the best way to do so is by sending your SQL statement to the CartoDB SQL API endpoint at http://[YOURUSERNAME].cartodb.com/api/v2/sql. This can be useful to debug issues in interaction with your web application in particular.

The browser string uses an encoded URL, which substitutes character sequences for some special characters. For example, you could use a URL encoder/decoder, which is easily found on the Web, to produce such a string.

Use the following instructions to see the result JSON returned by CartoDB given a particular SQL query. The URL string is also contained in c3/data/original/ url_query1.txt.

1. Enter the following URL string into your browser, substituting [YOURUSERNAME] with your CartoDB user name and [YOURAPIKEY] with your API key:

```
http://[YOURUSERNAME].cartodb.com/api/v2/sql?q=%20SELECT%20
toxic_channels.*%20FROM%20toxic_channels%20INNER%20
JOIN%20channels%20ON%20toxic_channels.join_BASIN%20=%20
channels.basin%20WHERE%20toxic_channels.join_order%20%3C%20
(SELECT%20channels._order%20FROM%20channels%20WHERE%20st_
distance(the_geom,%20ST_GeomFromText%20(%27POINT(-75.56111%20
39.72583)%27,4326))%20IN%20(SELECT%20MIN(st_distance(the_geom,%20
ST_GeomFromText(%27POINT(-75.56111%2039.72583)%27,4326)))%20
FROM%20channels%20x))%20AND%20toxic_channels.join_basin%20
=%20(SELECT%20channels.basin%20FROM%20channels%20WHERE%20st_
distance(the_geom,%20ST_GeomFromText%20(%27POINT(-75.56111%20
39.72583)%27,4326))%20IN%20(SELECT%20MIN(st_distance(the_geom,%20
ST_GeomFromText(%27POINT(-75.56111%2039.72583)%27,4326)))%20
FROM%20channels%20x))%20GROUP%20BY%20toxic_channels.cartodb_id%20
&api_key=[YOURAPIKEY]
```

2. Submit the browser request.

3. You will see a result similar to the following:

```
{"rows":[{"the_geom":"0101000020E610000056099A6A64E352C0B23A9C
05D4E84340","id":13,"join_segme":1786,"join_node_":1897,"join_
nod_1":1886,"join_basin":98,"join_order":2,"join_ord_1":6,"join_
lengt":1890.6533221,"distance":150.739169156001,"cartodb_
id":14,"created_at":"2015-05-06T21:52:52Z","updated_at":"2015-
05-06T21:52:52Z","the_geom_webmercator":"0101000020110F00000B1E9
E3DB30A60C18DCB53943D765241"},{"the_geom":"0101000020E610000011
44805EA1E652C0ECE7F94B65E64340","id":3,"join_segme":1710,"join_
node_":1819,"join_nod_1":1841,"join_basin":98,"join_
order":1,"join_ord_1":5,"join_lengt":769.46323073,"distan
ce":50.1031572450681,"cartodb_id":4,"created_at":"2015-05-
06T21:52:52Z","updated_at":"2015-05-06T21:52:52Z","the_geom_webme
rcator":"0101000020110F0000181D4045730D60C1F35490178D735241"},{"t
he_geom":"0101000020E61000009449A70ACFF052C0F3916D0D41D34340","id
":17,"join_segme":1098,"join_node_":1188,"join_nod_1":1191,"join_
basin":98,"join_order":1,"join_ord_1":5,"join_lengt":1320.8328273
,"distance":260.02935238833,"cartodb_id":18,"created_at":"2015-05-
06T21:52:52Z","updated_at":"2015-05-06T21:52:52Z","the_geom_webme
```

rcator":"0101000020110F00008167DA44181660C117FA8EFC695E5241"},{"t
he_geom":"0101000020E6100000DD53F65225EA52C0966E1B86B1E64340","id
":19,"join_segme":1728,"join_node_":1839,"join_nod_1":1826,"join_
basin":98,"join_order":1,"join_ord_1":5,"join_lengt":489.2571289,
"distance":201.8453893386,"cartodb_id":20,"created_at":"2015-05-
06T21:52:52Z","updated_at":"2015-05-06T21:52:52Z","the_geom_webme
rcator":"0101000020110F00009D303E9A6F1060C1BAD6D85BE1735241"},{"t
he_geom":"0101000020E61000008868F447FAE452C02218260DC0E94340","id
":12,"join_segme":1801,"join_node_":1913,"join_nod_1":1899,"join_
basin":98,"join_order":2,"join_ord_1":6,"join_lengt":539.82994246,
"distance":232.424790511141,"cartodb_id":13,"created_at":"2015-05-
06T21:52:52Z","updated_at":"2015-05-06T21:52:52Z","the_geom_webme
rcator":"0101000020110F00003BC511F10B0C60C1D801919542775241"},{"t
he_geom":"0101000020E6100000A2EE318E20EF52C0A874919E9BD44340","id
":16,"join_segme":1151,"join_node_":1243,"join_nod_1":1195,"join_
basin":98,"join_order":1,"join_ord_1":5,"join_lengt":1585.6022332,
"distance":48.7125304167275,"cartodb_id":17,"created_at":"2015-05-
06T21:52:52Z","updated_at":"2015-05-06T21:52:52Z","the_geom_webme
rcator":"0101000020110F000055062CA8AA1460C19A29734CE85F5241"},{"t
he_geom":"0101000020E610000043356AB28DEE52C090391E3073DF4340","id
":21,"join_segme":1548,"join_node_":1650,"join_nod_1":1633,"join_
basin":98,"join_order":3,"join_ord_1":7,"join_lengt":893.68816603
,"distance":733.948566072529,"cartodb_id":22,"created_at":"2015-
05-06T21:52:52Z","updated_at":"2015-05-06T21:52:52Z","the_geom_web
mercator":"0101000020110F0000F46510EE2D1460C18C0E2241E06B5241"},{
"the_geom":"0101000020E61000009B543F2277EA52C0F3615A0BD1D54340","i
d":1,"join_segme":1198,"join_node_":1292,"join_nod_1":1293,"join_
basin":98,"join_order":1,"join_ord_1":5,"join_lengt":746.7496066,
"distance":123.258432999702,"cartodb_id":2,"created_at":"2015-05-
06T21:52:52Z","updated_at":"2015-05-06T21:52:52Z","the_geom_webme
rcator":"0101000020110F0000CFB06115B51060C1B715F2AF3D615241"},{"t
he_geom":"0101000020E610000056AEF2E2D0EE52C0305E947734D94340","id
":9,"join_segme":1336,"join_node_":1432,"join_nod_1":1391,"join_
basin":98,"join_order":1,"join_ord_1":5,"join_lengt":1143.9037155
,"distance":281.665088681164,"cartodb_id":10,"created_at":"2015-
05-06T21:52:52Z","updated_at":"2015-05-06T21:52:52Z","the_geom_we
bmercator":"0101000020110F0000D8727BFE661460C1F269C8F6FA6452
41"}],"time":0.029,"fields":{"the_geom":{"type":"geometry"},
"id":{"type":"number"},"join_segme":{"type":"number"},"join_
node_":{"type":"number"},"join_nod_1":{"type":"number"},"join_
basin":{"type":"number"},"join_order":{"type":"number"},"join_
ord_1":{"type":"number"},"join_lengt":{"type":"number"},"distan
ce":{"type":"number"},"cartodb_id":{"type":"number"},"created_
at":{"type":"date"},"updated_at":{"type":"date"},"the_geom_webmerc
ator":{"type":"geometry"}},"total_rows":9}

Leaflet and an external API: CartoDB SQL

In the first section, we created a web application using Leaflet and the local GeoJSON files containing our layers. In this section, we will use Leaflet to display data from an external API—CartoDB SQL API. Perform the following steps:

1. Open the **qgis2leaf Export** dialog (navigate to **Web | qgis2leaf | Exports**).

2. In the **qgis2leaf** dialog, you can leave the inputs as their default ones. We will be heavily modifying the output code, so this part isn't so important. You may wish to add a basemap; MapQuest Open OSM is a good choice for this.

3. Take note of the output location.

4. Click on **OK**.

5. Locate index.html in the output directory.

6. Replace the contents of index.html with the following code (also available at c3/data/web/index.html). This code is identical to the existing index.html with a few modifications. All lines after 25 and the ones below the closing script, body, and HTML tags have been removed. The getToxic function and call have been added. Look for this function to replace the existing filler text with your CartoDB account name and API key. This function carries out our CartoDB SQL query and displays the results. We will comment out a second function call, which you may want to test to see the varying results based on the different coordinate pairs passed, as follows:

```
<!<!<!DOCTYPE html>>>>
<html>
  <head>
    <title>QGIS2leaf webmap</title>

    <meta charset="utf-8" />
    <link rel="stylesheet" href="http://cdnjs.
      cloudflare.com/ajax/libs/leaflet/0.7.3/
      leaflet.css" /> <!-- we will use this as the
      styling script for our webmap-->
    <link rel="stylesheet" href="css/MarkerCluster.css"
      />
    <link rel="stylesheet" href="css/Marker
      Cluster.Default.css" />
    <link rel="stylesheet" type="text/css"
      href="css/own_style.css"/>
    <link rel="stylesheet" href="css/label.css" />
    <script src="http://code.jquery.com/jquery-
      1.11.1.min.js"></script> <!-- this is the javascript
      file that does the magic-->
```

```
    <script src="js/Autolinker.min.js"></script>
</head>
<body>
    <div id="map"></div> <!-- this is the initial look of
       the map. in most cases it is done externally using
       something like a map.css stylesheet where you can
       specify the look of map elements, like background
       color tables and so on.-->
    <script src="http://cdnjs.cloudflare.com/ajax/
       libs/leaflet/0.7.3/leaflet.js"></script> <!-- this
       is the javascript file that does the magic-->
    <script src="js/leaflet-hash.js"></script>
    <script src="js/label.js"></script>
    <script src="js/leaflet.markercluster.js"></script>


    <script src='data/exp_toxicchannels.js' ></script>


    <script>
      var map = L.map('map', { zoomControl:true
        }).fitBounds([[39.4194805496,-75.8685268698]
        ,[39.9951967581,-75.2662748017]]);
      var hash = new L.Hash(map); //add hashes to html
        address to easy share locations
      var additional_attrib = 'created w. <a
        href="https://github.com/geolicious/qgis2leaf"
        target ="_blank">qgis2leaf</a> by <a
        href="http://www.geolicious.de" target
        ="_blank">Geolicious</a> & contributors<br>';
      var feature_group = new L.featureGroup([]);


      var raster_group = new L.LayerGroup([]);


      var basemap_0 = L.tileLayer('http://otile1.
        mqcdn.com/tiles/1.0.0/map/{z}/{x}/{y}.jpeg', {
          attribution: additional_attrib + 'Tiles Courtesy
          of <a href="http://www.mapquest.com/">MapQuest
          </a> &mdash; Map data: &copy; <a href="
          http://openstreetmap.org">OpenStreetMap</a>
          contributors,<a href="http://creativecommons.org
          /licenses/by-sa/2.0/">CC-BY-SA</a>'});
      basemap_0.addTo(map);
      var layerOrder=new Array();
      function pop_toxicchannels(feature, layer) {
```

```
var popupContent = '<table><tr><th
  scope="row">ID</th><td>' +
  Autolinker.link(String(feature.properties['ID']))
  + '</td></tr><tr><th
  scope="row">join_SEGME</th><td>' +
  Autolinker.link(String(feature.properties
  ['join_SEGME'])) + '</td></tr><tr><th
  scope="row">join_NODE_</th><td>' +
  Autolinker.link(String(feature.properties
  ['join_NODE_'])) + '</td></tr><tr><th
  scope="row">join_NOD_1</th><td>' +
  Autolinker.link(String(feature.properties
  ['join_NOD_1'])) + '</td></tr><tr><th
  scope="row">join_BASIN</th><td>' +
  Autolinker.link(String(feature.properties
  ['join_BASIN'])) + '</td></tr><tr><th
  scope="row">join_ORDER</th><td>' +
  Autolinker.link(String(feature.properties
  ['join_ORDER'])) + '</td></tr><tr><th
  scope="row">join_ORD_1</th><td>' +
  Autolinker.link(String(feature.properties
  ['join_ORD_1'])) + '</td></tr><tr><th
  scope="row">join_LENGT</th><td>' +
  Autolinker.link(String(feature.properties
  ['join_LENGT'])) + '</td></tr><tr><th
  scope="row">distance</th><td>' +
  Autolinker.link(String(feature.properties
  ['distance'])) + '</td></tr></table>';
layer.bindPopup(popupContent);



}


var exp_toxicchannelsJSON = new L.geoJson
  (exp_toxicchannels,{
  onEachFeature: pop_toxicchannels,
  pointToLayer: function (feature, latlng) {
    return L.circleMarker(latlng, {
      radius: feature.properties.radius_qgis2leaf,
      fillColor: feature.properties.color_qgis2leaf,

      color: feature.properties.borderColor
        _qgis2leaf,
      weight: 1,
      opacity: feature.properties.transp_qgis2leaf,
      fillOpacity: feature.properties.
        transp_qgis2leaf
    })
```

```
    }
});
feature_group.addLayer(exp_toxicchannelsJSON);

layerOrder[layerOrder.length] = exp_toxic
  channelsJSON;
for (index = 0; index < layerOrder.length; index++) {
  feature_group.removeLayer(layerOrder[index]);
  feature_group.addLayer(layerOrder[index]);
}

//add comment sign to hide this layer on the map in
  the initial view.
exp_toxicchannelsJSON.addTo(map);
var title = new L.Control();
title.onAdd = function (map) {
  this._div = L.DomUtil.create('div', 'info'); //
    create a div with a class "info"
  this.update();
  return this._div;
};
title.update = function () {
  this._div.innerHTML = '<h2>This is the title</h2>
    This is the subtitle'
};
title.addTo(map);
var baseMaps = {
  'MapQuestOpen OSM': basemap_0
};
L.control.layers(baseMaps,{"toxicchannels":
  exp_toxicchannelsJSON},{collapsed:false})
  .addTo(map);
L.control.scale({options: {position:
  'bottomleft',maxWidth: 100,metric: true,imperial:
  false,updateWhenIdle: false}}).addTo(map);

/* we've inserted the following after the existing
  index.html line 83, to handle query to cartodb */

function getToxic(lon,lat)
{
  var toxicLayer = new L.GeoJSON();

    $.getJSON(
```

```
"http://YOURCARTODBACCOUNTNAMEHERE.cartodb.com/api/v2/sql?q=%20
SELECT%20toxic_channels.*%20FROM%20toxic_channels%20INNER%20
JOIN%20channels%20ON%20toxic_channels.join_BASIN%20=%20channels.
basin%20WHERE%20toxic_channels.join_order%20%3C%20(SELECT%20
channels._order%20FROM%20channels%20WHERE%20st_distance(the_
geom,%20ST_GeomFromText%20(%27POINT(" + lon + "%20" + lat +
")%27,4326))%20IN%20(SELECT%20MIN(st_distance(the_geom,%20ST_
GeomFromText(%27POINT(" + lon + "%20" + lat + ")%27,4326)))%20
FROM%20channels%20x))%20AND%20toxic_channels.join_basin%20
=%20(SELECT%20channels.basin%20FROM%20channels%20WHERE%20st_
distance(the_geom,%20ST_GeomFromText%20(%27POINT(" + lon + "%20" +
lat + ")%27,4326))%20IN%20(SELECT%20MIN(st_distance(the_geom,%20
ST_GeomFromText(%27POINT(" + lon + "%20" + lat + ")%27,4326)))%20
FROM%20channels%20x))%20GROUP%20BY%20toxic_channels.cartodb_id%20
&api_key=YOURCARTODBAPIKEYHERE&format=geojson&callback=?",
            function(geojson) {
               $.each(geojson.features, function(i, feature)
                 {
                 toxicLayer.addData(feature);
               })
            });
            map.addLayer(toxicLayer);
         }

      getToxic(-75.56111,39.72583);

      //getToxic(-75.70993,39.69099);

   </script>
  </body>
</html>
```

Your results should look similar to the following image:

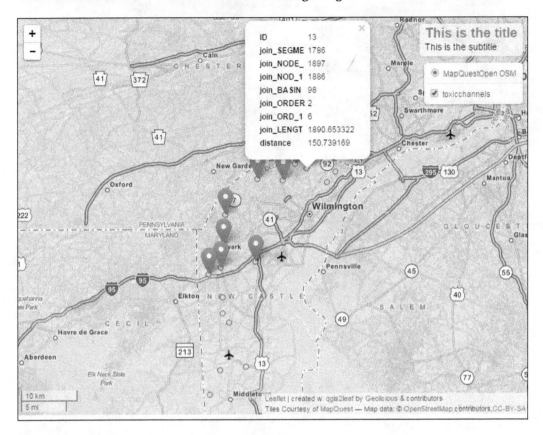

Summary

In this chapter, we produced a dynamic web application using a physical raster analysis example: hydrological analysis. To do this, we started by preparing the raster elevation data for the hydrological analysis and then performed the analysis. We took a look at how we could automate that workflow using the Modeler workflow automation tool. Next, we used NNJoin to create a spatial join between some hydrological outputs to produce a data source that would be suited to web interaction and querying. Finally, we published this data to an external cloud platform, CartoDB, and implemented their SQL API in a JavaScript function to find the toxic sites upstream from a location, given the Leaflet web client interaction. In the next chapter, we will produce a web application using network analysis and crowd sourced interaction.

4
Finding the Best Way
to Get There

In this chapter, we will explore formal network-like geographic vector object relationships. Topological relationships are useful in many ways for geographical data management and analysis, but perhaps the most important application is optimal path finding. Specifically, you will learn how to make a few visualizations related to optimal paths: isochron polygons and accumulated traffic lines. With these visual elements as a background, we will incorporate social media feedback through Twitter in our web map application. The end result will be an application that communicates back and forth with the stakeholders about safe school routes.

In this chapter, we will cover the following topics:

- Downloading OpenStreetMap data
- Spatial queries
- Installing Postgres/PostGIS/pgRouting
- Building a topological network
- DB Manager
- Using the shortest path plugin to test the topology
- Generating the costs to travel to a point for each road segment
- Creating the isochron contours
- Generating the shortest paths for all students
- Adding Twitter data through Python

Postgres with PostGIS and pgRouting

Vector-based GIS, if not by definition then de facto, are organized around databases of geographic objects, storing their geometric definitions, geographic metadata, object relationships, and other attributes. Postgres is a leading open source relational database platform. Unlike SQLite, this is not a file-based system, but rather it requires a running service on an available machine, such as the localhost or an accessible server. The spatial extension to Postgres, PostGIS, provides all the functionalities around geospatial data, such as spatial references, geographic transformation, spatial relationships, and more. Most recently, PostGIS has come to support topology — the formal relationships between geometric objects. pgRouting is a topological analysis engine built around optimal path-finding. Conveniently, PostGIS now comes bundled with pgRouting. The following content applies to Postgres 9.3.

Installing Postgres/PostGIS/pgRouting

On Windows, you can use the Postgres installer to install PostGIS and pgRouting along with Postgres. On Mac, you can use the Kyngchaos binary installer found at `http://www.kyngchaos.com/software/postgres`. On Linux, you can refer to the PostGIS installation documentation for your distribution found at `http://postgis.net/install/`.

The installation instructions for Windows are as follows:

1. Download the Postgres installer from `http://www.postgresql.org/download/windows/` and start it.
2. Follow the prompt; pick a password.
3. Click on **Launch Stack Builder at exit** and then click on **Finish**.
4. In **Stack Builder**, select the PostgreSQL instance you just created.
5. Check **PostGIS 2.1** Bundle under **Spatial Extensions**.
6. Select **PostGIS** and **Create spatial database** under the **Choose Components** dialog.
7. Finish the installation, skipping the database creation steps, which are prone to failure.

Creating a new Postgres database

Now that we installed the Postgres database server with PostGIS, the pgRouting extensions, and the pgAdmin III client program, we want to create a new database where we can work. Perform the following steps:

1. Open the pgAdmin III program.

2. Right-click on the **Databases** section of the **Servers** tree and click on **New Database...** to create the database, as shown in the following screenshot:

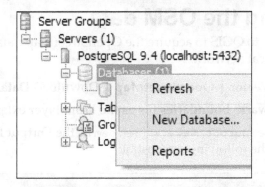

3. Enter a name for the new database, `packt`, and click on **OK**.

Registering the PostGIS and pgRouting extensions

Next, we need to tell Postgres that we want to use the PostGIS and pgRouting extensions with our new database. Perform the following steps:

1. Open pgAdmin III if you haven't already done so.

2. Navigate to **Tools | Query**.

3. Enter the following SQL in the **SQL Editor** area:

```
CREATE EXTENSION postgis;
CREATE EXTENSION pgrouting;
```

4. From the **Query** menu, choose **Execute**.

OpenStreetMap data for topology

A topological network, which specifies the formal relationships between geometric objects, requires real geographic data for it to be useful in an actual physical space. So next, we will acquire some geographic data in order to construct a network providing the shortest path between points in a physical space, following certain rules embedded in the network. A great source of data for this, and many other purposes, is OpenStreetMap.

Downloading the OSM data

Now, let's move back to QGIS to acquire the OpenStreetMap data from which we will create a topological network:

1. Navigate to **Vector | OpenStreetMap | Download Data**.

2. Select the **newark_boundaries** file as the **From layer** extent.

3. Enter c4/data/output/newark_osm.osm as the **Output file** and click on **OK**, as shown in the following screenshot:

Adding the data to the map

The downloaded data must be added to the QGIS project to verify that it has been downloaded and to further work on the data from within QGIS. Perform the following steps:

1. Navigate to **Layer | Add Layer | Add Vector Layer**.

2. Select `c4/data/output/newark_osm.osm` as the **Source**.

3. Click on **Select All** from the **Select vector layers...** dialog.

4. Click on **OK**.

5. You should now see the data displayed, looking similar to the following image:

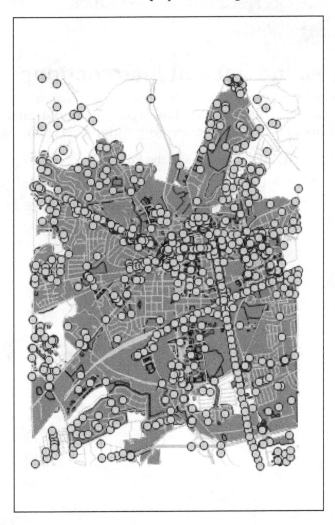

Projecting the OSM data

We will project the OSM data onto the projection used by the other data to be added to the project, which is the location of the students. We want these two datasets to use the same projection system; otherwise, we will run into trouble while building our topological network and analyzing the network. Perform the following steps:

1. Select the **lines** layer from the **Layers** panel.

2. Go to **Layer | Save as**.

3. Enter the following parameters:

    1. **Save as**: `c4/data/output/newark_osm.shp`.

    2. Select **CRS EPSG:2880** (Delaware Ft/HARN).

    3. Click on **OK**.

Splitting all the lines at intersections

It is necessary that the topological edges to be created are coterminous with the geographic data vertices. This is called a **topologically correct** dataset. We will use **Split lines with lines** to fulfill this requirement. Perform the following steps:

1. Search for **Split lines with lines** in the **Processing Toolbox** panel.

2. Select the projected OSM lines file, `c4/data/output/newark_osm.shp`, as both the **Input layer** and **Split layer**.

3. Click on **Run**, as shown in the following screenshot:

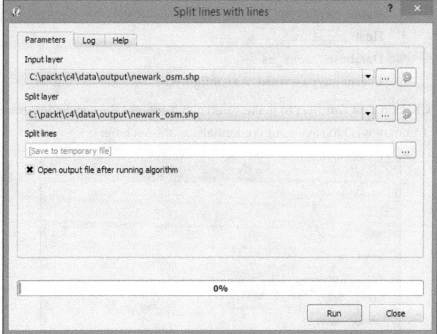

Database importing and topological relationships

Now that we've prepared the OSM data, we need to actually load it into the database. Here, we can generate the topological relationships based on geographic relationships as determined by PostGIS.

Connecting to the database

Although we will be working from the Database Manager when dealing with the database in QGIS, we will first need to connect to the database through the normal "Add Layer" dialog. Perform the following steps:

1. Navigate to **Layer | Add Layer | Add PostGIS Layers**.
2. Click on **New**.

3. In the **Create a New PostGIS connection** dialog, enter the following parameters, accepting others as their defaults:

   ◦ **Name**: `packt_c4`

   ◦ **Host**: `localhost`

   ◦ **Database**: `packt_c4`

   ◦ **Username/password**: As configured earlier in this chapter

4. Click on **Test Connect** to make sure you've entered the correct information.

5. You may wish to save your credentials, as shown here:

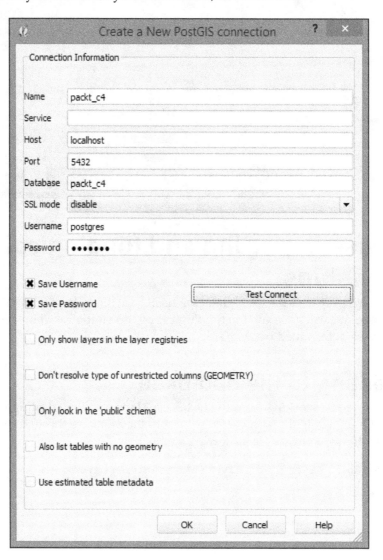

Importing into PostGIS with DB Manager

Once we've added the database connection, DB Manager is where we'll be interacting with the database. DB Manager provides query access via the SQL syntax as well as the facility to add results as a virtual (in memory, not on disk) layer. We can also use DB Manager to import or export data to/from the database when necessary. Perform the following steps:

1. Go to **Database | DB Manager | DB Manager**.

2. You may need to navigate to **Database | Refresh** to have a new database appear.

3. Select the database to be updated (for example, `packt_c4`). The following is an image of the Database Manager and tables, which were generated when you created your new PostGIS database:

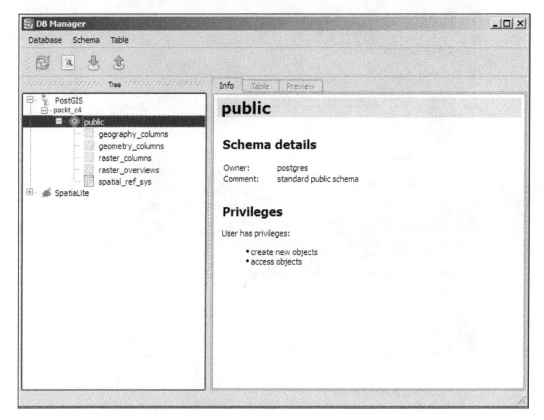

4. Navigate to **Table | Import layer/file**.

5. Input the following parameters:
   - ° **Input: Split lines**
   - ° **Table: newark_osm**
   - ° **Source SRID**: 2880
   - ° **Target SRID**: 2880
   - ° Select **Create single-part geometries instead of multi-part**
   - ° Select **Create spatial index**, as shown in the following screenshot:

Repeat these steps with the **students** layer:

1. Add the **students** layer from `c4/data/original/students.shp`.

2. Repeat the previous **Import vector layer** steps with the **students** layer, as shown in the following screenshot:

The imported tables and the associated schema and metadata information will now be visible in DB Manager, as shown in the following screenshot:

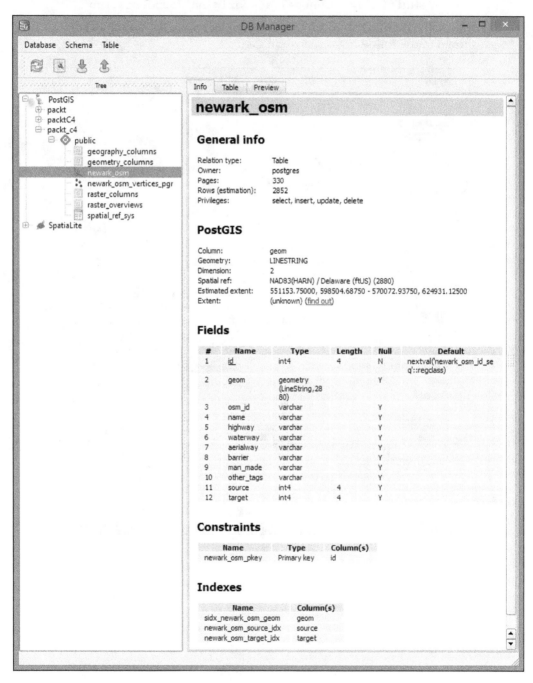

Creating the topological network data

Next, run a query that adds the necessary fields to the newark_osm table, updating these with the topological information, and create the related table of the network vertices, newark_osm_vertices. These field names and types, expected by pgRouting, are added by the alter queries and populated by the pgr_createTopology pgRouting function. The length_m field is populated with the segment length using an update query with the st_length function (and st_transform here to control the spatial reference). This field will be used to help determine the cost of the shortest path (minimum cost) routing. Perform the following steps:

1. Navigate to **Database | DB Manager | DB Manager**.

2. Select the database to be updated.

3. Go to **Database | SQL window**. Enter the following code:

```
alter table newark_osm add column source integer;
alter table newark_osm add column target integer;
select pgr_createTopology('newark_osm', 0.0001, 'geom',
    'id');

alter table newark_osm add column length_m float8;
update newark_osm set length_m = st_length
    (st_transform(geom,2880));
```

An alternate workflow: topology with osm2po

The osm2po program performs many topological dataset preparation tasks that might otherwise require a longer workflow — such as the preceding task. As the name indicates, it is specifically used for the OpenStreetMap data. The osm2po program must be downloaded and installed separately from the osm2po website, http://osm2po.de. Once the program is installed, it is used as follows:

```
[..] > cd c:\packt\c4\data\output
c:\packt\c4\data\output>java -jar osm2po-5.0.0\osm2po-core-5.0.0-
    signed.jar cmd=tj
sp newark_osm.osm
```

This command will create a .sql file that you can run in your database to add the topological table to your database, producing something very similar to what we did in the preceding section.

Using the pgRouting Layer plugin to test

Let's use the pgRouting Layer plugin to test whether the steps we've performed up to this point have produced a functioning topological network to find the shortest path. We will find the shortest path between two arbitrary points on the network: 1 and 1000. Perform the following steps:

1. Install the pgRoutingLayer plugin.
2. If the shortest path panel is not displayed, turn it on under **View | Panels | pgRouting Layer**.
3. Enter the following parameters:
   - **Database: packt_c4**.
   - Ensure that you are already connected to the database, as shown in the previous section. You may need to restart QGIS for a new database connection to show up here.
   - **edge_table**: newark_osm
   - **geometry**: geom
   - **id**: id
   - **source**: source
   - **target**: target
   - **cost**: length_m
   - **source_id**: 1
   - **target_id**: 1000

Your output will look similar to the following image:

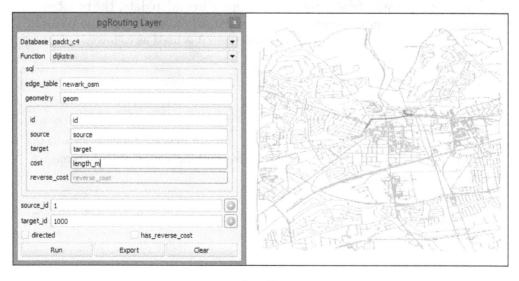

Creating the travel time isochron polygons

Let's say that the school in our study is located at the vertex with an ID of 1 in the `newark_osm` layer. To visualize the walking time from the students' homes, without releasing sensitive information about where the students actually live, we can create isochron polygons. Each polygon will cover the area that a person can walk from to a single destination within some time threshold.

Generating the travel time for each road segment

We'll use DB Manager to create and populate a column for the travel time on each segment at the walking speed; then, we will create a query layer that includes the travel time from each road segment to our school at vertex 1. Perform the following steps:

1. Navigate to **Database | DB Manager | DB Manager**.

2. Select the database to be updated.

3. Go to **Database | SQL window**.

4. Enter the following code:

```
ALTER TABLE newark_osm ADD COLUMN traveltime_min float8;
UPDATE newark_osm SET traveltime_min = length_m  / 6000.0 *
  60;

SELECT *
FROM pgr_drivingdistance('SELECT id, source, target,
  traveltime_min as cost FROM newark_osm'::text, 1,
  100000::double precision, false, false) di (seq, id1,
  id2, cost)
JOIN newark_osm rd ON di.id2 = rd.id;
```

5. Select the **Load as new layer** option.

6. Select **Retrieve columns**.

7. Select **seq** as your **Column with unique integer values** and **geom** as your **Geometry column**.

8. Click on the **Load now!** button, as shown in the following screenshot:

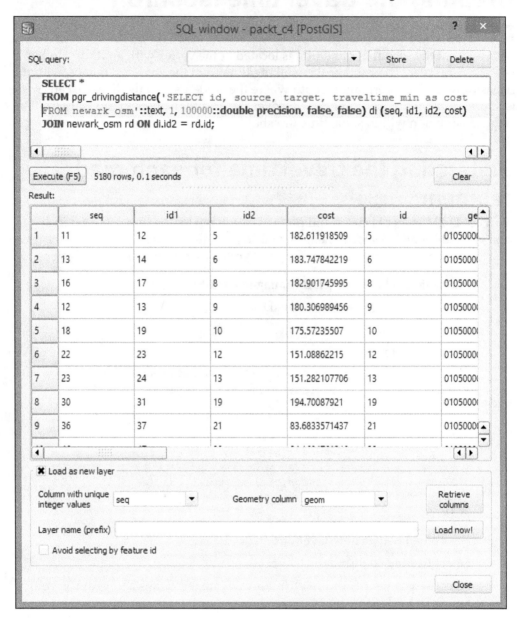

You can now symbolize the segments by the time it takes to get from that location to the school. To do this, use a **Graduated** style type with the `traveltime_min` field. You will see that the network segments with lower values (indicating quicker travel) are closer to vertex 1, and the opposite is true for the network segments with higher values. This method is limited by the extent to which the network models real conditions; for example, railroads are visualized along with other road segments for the travel time. However, railroads could cause discontinuity in our network—as they are not "traversable" by students traveling to school.

Creating isochron polygons

Next, we will create the polygons to visualize the areas from which the students can walk to school in certain time ranges. We can use this technique to characterize the general travel time and keep the student locations hidden.

Converting the travel time lines to points

We will need to first convert our current line-based travel time layer to points (centroids), using the polygons as an intermediate step. Perform the following steps:

1. Save the query layer as a shapefile: `c4/data/output/newark_isochrone.shp`.

2. Navigate to **Vector | Geometry Tools | Line to polygons**. Input the following parameters:

    ° **Input line vector layer**: isochron lines

    ° **Output polygon shapefile**: `c4/data/output/isochron_polygon.shp`

    ° Click on **OK**

3. Navigate to **Vector | Geometry Tools | Polygons to centroid**. Input the following parameters:

    ° **Input polygon vector layer**: `c4/data/output/isochron_polygon.shp`

    ° **Output point shapefile**: `c4/data/output/isochrons_centroids.shp`

    ° Click on **OK**

Selecting the travel time ranges in points and creating convex hulls

Next, we'll create the actual isochron polygons for each time bin. We must select each set of travel time points using a filter expression for the three time periods: 15 minutes or less, 30 minutes or less, and 45 minutes or less. Then, we'll run the **Concave hull** tool on each selection. This will create a polygon feature around each set of points.

You'll perform the following steps three times for each of the three break values, which are 15, 30, and 45:

1. Select **isochron_centroids** from the **Layers** panel.

2. Navigate to **Layer | Query**.

3. Click on **Clear** if there is already a filter expression displayed in the filter expression field of the query dialog.

4. Provide a specific field expression: `cost < [break value]` (for example, `cost < 15`).

5. Click on **OK** to select the objects in the layer that matches the expression.

6. Navigate to **Processing Toolbox | Concave hull**.

7. Input the following parameters for **Concave hull**. All other parameters can be left at their defaults:

   ° **Input point layer: isochron_centroids**

   ° Select **Split multipart geometry into singleparts geometries**

   ° **Concave hull** (the output file) could be similar to `c4/data/output/isochron45.shp`

   ° Click on **Run**, as shown in the following screenshot:

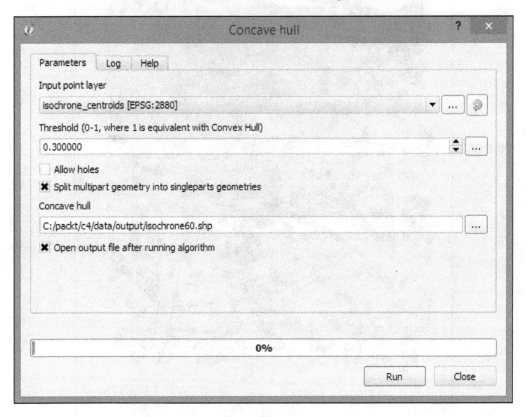

All concave hulls when displayed will look similar to the following image. The "spikiness" of the concave hulls reflects relatively few road segments (points) used to calculate these travel time polygons:

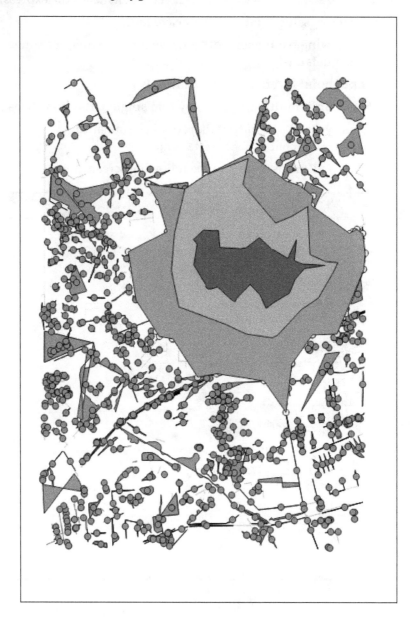

Generating the shortest paths for all students

So far, we have only looked at the shortest path between all the given segments of road in the city. Now, given the student location, let's look at where student traffic will accumulate.

Finding the associated segment for a student location

By following these steps, we will join attributes from the closest road segment — including the associated topological and travel attributes — to each student location. Perform the following steps:

1. Install the NNJoin plugin.

2. Navigate to **Plugins | NNJoin | NNJoin**.

3. Enter the following parameters:

    ○   **Input vector layer: students**

    ○   **Join vector layer: newark_osm**

    ○   **Output layer**: students_topology

4. Click on **OK**.

5. Import students_topology into the packt_c4 database using Database Manager.

The following image shows the parameters as entered into the NNJoin plugin:

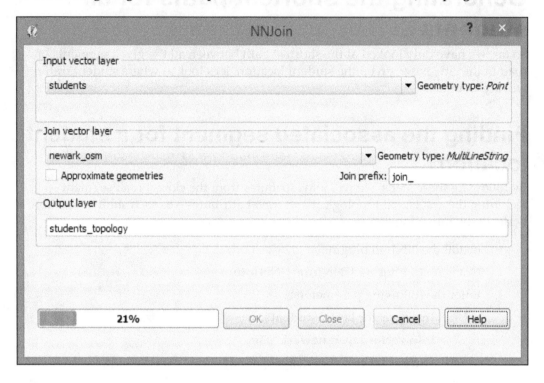

Calculating the accumulated shortest paths by segment

We want to find which routes are the most popular given the student locations, network characteristics, and school location. The following steps will produce an accumulated count of the student traffic along each network segment:

1. Go to **Database | DB Manager | DB Manager**.

2. Select the database to be updated.

3. Navigate to **Database | SQL window**.

4. We want to run a SQL command that will do a shortest path calculation for each student and find the total number of students traveling on each road segment. This query may be very slow. Enter the following code:

```
SELECT id, geom, count(id1)
FROM
(SELECT *
  FROM pgr_kdijkstraPath(
```

```
        'SELECT id, source, target, traveltime_min as cost FROM
newark_osm',
        1, (SELECT array_agg(join_target) FROM students_topology),
false, false
        ) a,
    newark_osm b
WHERE a.id3=b.id) x
GROUP BY id, geom
```

5. Select the **Load as new layer** option.

6. Select **id** as your **Column with unique integer values** and **geom** as your **Geometry column**, as shown in the following screenshot:

Flow symbology

We want to visualize the number of students on each segment in a way that really accentuates the segments that have a high number of students traveling on them. A great way to do this is with a symbology expression. This produces a graduated symbol as would be found in other GIS packages. Perform the following steps:

1. Navigate to **Layer | Properties | Style**.

2. Click on **Simple line** to access the symbology expressions, as shown in the following screenshot:

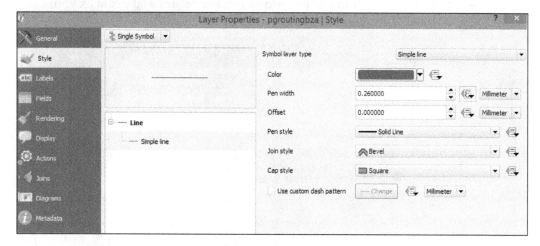

3. In the **Pen width** section, click on the advanced menu to edit the symbology expression.

4. Natural log is a good function to use to get a more linear growth rate when a value grows exponentially. This helps us to produce a symbology that varies in a more visually appealing way. Enter the following expression into the **Expression** string builder dialog:

    ```
    ln("count")
    ```

Now that we have mapped the variable sized symbol to the natural log of the count of students traveling on each segment, we will see a pleasing visualization of the "flow" of students traveling on each road segment. The student layer, showing the student locations, is displayed alongside the flow to better illustrate what the flow visualization shows.

Web applications – creating safe corridors

Decision makers can use the accumulated shortest paths output to identify the busiest paths to the school. They can use this information to communicate with guardians about the safest routes for their children.

Planners can go a step further by investing in safe infrastructure for the most used paths. For example, planners can identify the busy crossings over highways using the "count" attribute (and visualization) from the query layer and the "highways" attribute from `newark_osm`.

Of course, communication with stakeholders is ideally a two-way process. To achieve this goal, planners could establish a social networking account, such as a Twitter account, for parents and students to report the problems or features of the walking routes. Planners would likely want to look at this data as well to adjust the safe routes to the problem spots or amenities. This highly simplistic model should be adjusted for the other variables that could also be captured in the data and modeled, such as high traffic roads and so on.

Registering a Twitter account and API access

The following instructions will direct you on how to set up a new Twitter account in a web browser and get your community to make geotagged tweets:

1. Create a Twitter account for this purpose (a nonpersonal one). The account will need to be linked to a unique mobile phone and e-mail. If you've already linked your e-mail and mobile phone with an account, there are some hints for getting around this in the following section.

2. Go to `https://apps.twitter.com/` and create a new app.

3. You will need to unregister your phone number if it is already registered with another account (Twitter will warn you about this).

4. In **Application Settings**, find **manage keys and access tokens**. Here, you will find your consumer key and secret.

5. You must also create an access token by clicking on **Create my Access Token**.

6. Users who would like to have their tweets added to the system should be directed to use your Twitter handle (`@YOURNAME`). Retweet the tweets that you wish to add to your map. For a more passive solution, you can also follow all the users who you wish to capture; although, you'll need to find some way to filter out their irrelevant tweets.

Setting up the Twitter Tools API

We must now download and install Python-based Twitter Tools, which leverage the Twitter API. This will allow us to pull down GeoJSON from our Twitter account. Perform the following steps:

1. Download the Twitter Tools API from GitHub: `https://github.com/ sixohsix/twitter`.

2. Open the OSGeo4W shell using the **Run as administrator** command via the context menu, or if you're on Mac or Linux, use `sudo` to run it with full privileges. Your OS Account must have administrator privileges. Navigate to `C:\Program Files\QGIS Wien\OSGeo4W.bat`.

3. Extract the Twitter Tools API code and change drive (`cd`) into the directory that you extracted into (for example, `C:\Users\[YOURUSERNAME]\ Downloads\twitter-master\twitter-master`), using the following command line:

   ```
   > cd C:\Users\[YOURUSERNAME]\Downloads\twitter-master\twitter-
   master
   ```

4. Run the following from the command line to install the Twitter Tools software and dependencies:

   ```
   > python setup.py install
   > twitter
   ```

 Running `python setup.py install` in a directory containing the `setup.py` file on a path including the Python executable is the normal way to build (install) a Python program. You will need to install the `setuptools` module beforehand. The instructions to do so can be found on this website: `https://pypi.python.org/pypi/setuptools`.

5. Accept the authorization for the command-line tools. You will need to copy and paste a PIN (as given) from the browser to the command line.

6. Exit the command-line shell and start another OSGeo4W shell under your regular account.

7. You can use `twitter --help` for more options. Execute the following in the command line:

   ```
   > twitter --format json 1> "C:\packt\c4\data\output\twitter.json"
   > cd c:\packt\c4
   > python
   ```

8. Run the following in the interpreter (refer to the following section to run it noninteractively):

```
import json
f = open('./output/twitter.json', 'r')
jsonStr = f.read()
f.close()
jpy = json.loads(jsonStr)
geojson = ''
for x in jpy['safe']:
  if x['geo'] :
    geojson += '{"type": "Feature","geometry": {"type":
      "Point", "coordinates": [' + str(x['geo']
      ['coordinates'][1]) + ',' + str(x['geo']
      ['coordinates'][0]) + ']}, "properties": {"id": "' +
      str(x['id']) + '", "text": "' + x['text'] + '"}},'
geojson = geojson[:-1]
geojson += ']}'
geojson = '{"type": "FeatureCollection","features": [' +
  geojson
f = open('./data/output/twitter.geojson', 'w')
f.write(geojson)
f.close()
```

Or run the following command:

```
> python twitterJson2GeoJson.py
```

Here is an example of the GeoJSON-formatted output:

```
{"type": "FeatureCollection","features": [{"type":
"Feature","geometry": {"type": "Point", "coordinates":
[-75.75451,39.67434]}, "properties": {"id": "6064543366212530177",
"text": "Hello world"}},{"type": "Feature","geometry": {"type":
"Point", "coordinates": [-75.73968,39.68139]}, "properties":
{"id": "606454626456473600", "text": "Testing"}},{"type":
"Feature","geometry": {"type": "Point", "coordinates":
[-75.76838,39.69243]}, "properties": {"id": "606479472271826944",
"text": "Test"}}]}
```

Save this as `c4/output/twitter.geojson` from a text editor and import the file into QGIS as a vector layer to preview it along with the other layers. When these layers are symbolized, you may see something similar to the following image:

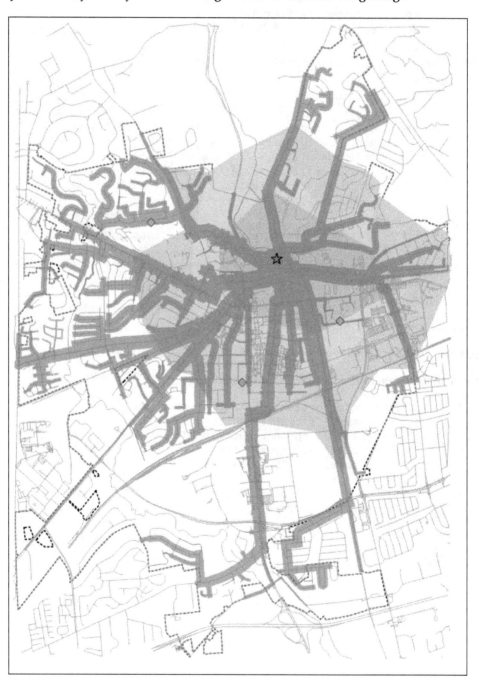

Finally, export the web application with qgis2leaf. You will notice some loss of information and symbology here. In addition, you may wish to customize the code to take advantage of the data and content passed through Twitter.

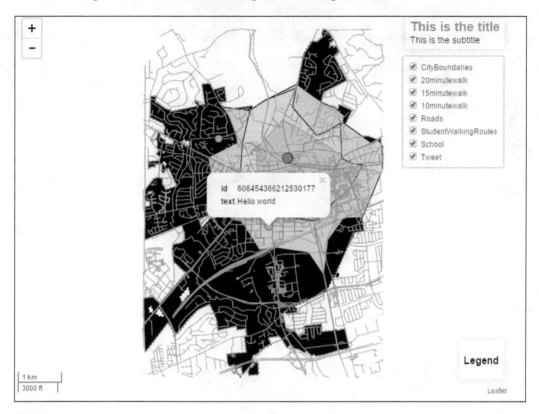

Summary

In this chapter, through a safe route selection example, we built a topological network using OSM data and Postgres with its PostGIS and pgRouting extensions. Using this network, we modeled the travel time to school from different locations on the road network and the students' travel to school, visualizing which routes were more and less frequently used. Finally, we added the contributed social network data on Twitter through a Python-based API, which we built using a typical Python build process. We then exported all the results using the same method as we did in the previous chapters: qgis2leaf. In the next chapter, we will explore the relationship between time and space and visualization through some new libraries.

5
Demonstrating Change

In this chapter, we will encounter the visualization and analytical techniques of exploring the relationships between place and time and between the places themselves.

The data derived from temporal and spatial relationships is useful in learning more about the geographic objects that we are studying—from hydrological features to population units. This is particularly true if the data is not directly available for the geographic object of interest: either for a particular variable, for a particular time, or at all.

In this example, we will look at the demographic data from the US Census applied to the State House Districts, for election purposes. Elected officials often want to understand how the neighborhoods in their jurisdictions are changing demographically. Are their constituents becoming younger or more affluent? Is unemployment rising? Demographic factors can be used to predict the issues that will be of interest to potential voters and thus may be used for promotional purposes by the campaigns.

In this chapter, we will cover the following topics:

- Using spatial relationships to leverage data
- Preparing data relationships for static production
- Vector simplification
- Using TopoJSON for vector data size reduction and performance
- D3 data visualization for API
- Animated time series maps

Leveraging spatial relationships

So far, we've looked at the methods of analysis that take advantage of the continuity of the gridded raster data or of the geometric formality of the topological network data.

For ordinary vector data, we need a more abstract method of analysis, which is establishing the formal relationships based on the conditions in the spatial arrangement of geometric objects.

For most of this section, we will gather and prepare the data in ways that will be familiar. When we get to preparing the boundary data, which is leveraging the State House Districts data from the census tracts, we will be covering new territory – using the spatial relationships to construct the data for a given geographic unit.

Gathering the data

First, we will gather data from the sections of the US Census website. Though this workflow will be particularly useful for those working with the US demographic data, it will also be instructive for those dealing with any kind of data linked to geographic boundaries.

To begin with, obtain the boundary data with a unique identifier. After doing this, obtain the tabular data with the same unique identifier and then join on the identifier.

Boundaries

Download 2014 TIGER/Line Census Tracts and State Congressional Districts from the US Census at `https://www.census.gov/geo/maps-data/data/tiger-line.html`.

1. Select **2014** from the tabs displayed; this should be the default year.

2. Click on the **Download** accordion heading and click on **Web interface**.

3. Under **Select a layer type**, select **Census Tracts** and click on **submit**; under **Census Tract**, select **Pennsylvania** and click on **Download**.

4. Use the back arrow if necessary to select **State Legislative Districts**, and click on **submit**; select **Pennsylvania** for **State Legislative Districts - Lower Chamber (current)** and click on **Download**.

5. Move both the directories to `c5/data/original` and extract them.

> We've only downloaded a single boundary dataset for this exercise. Since the boundaries are not consistent every year, you would want to download and work further with each separate annual boundary file in an actual project.

Tabular data from American FactFinder

Many different demographic datasets are available on the American FactFinder site. These complement the TIGER/Line data mentioned before with the attribute data for the TIGER/Line geographic boundaries. The main trick is to select the matching geographic boundary level and extent between the attribute and the geographic boundary data. Perform the following steps:

1. Go to the US Census American FactFinder site at `http://factfinder.census.gov`.

2. Click on the **ADVANCED SEARCH** tab.

3. In the **topic or table name** input, enter `White` and select **B02008: WHITE ALONE OR IN COMBINATION WITH ONE OR MORE RACES** in the suggested options. Then, click on **GO**.

4. From the sidebar, in the **Select a geographic type:** dropdown in the **Geographies** section, select **Census Tract - 140**.

5. Under **select a state**, select **Pennsylvania**; under **Select a county**, select **Philadelphia**; and under **Select one or more geographic areas and click Add to Your Selections:**, select **All Census Tracts within Philadelphia County, Pennsylvania**. Then, click on **ADD TO YOUR SELECTIONS**.

6. From the sidebar, go to the **Topics** section. Here, in the **Select Topics to add to 'Your Selections'** under **Year**, click on each year available from **2009** to **2013**, adding each to **Your Selections** to be then downloaded.

7. Check each of the five datasets offered under the **Search Results** tab. All checked datasets are added to the selection to be downloaded, as shown in the following screenshot:

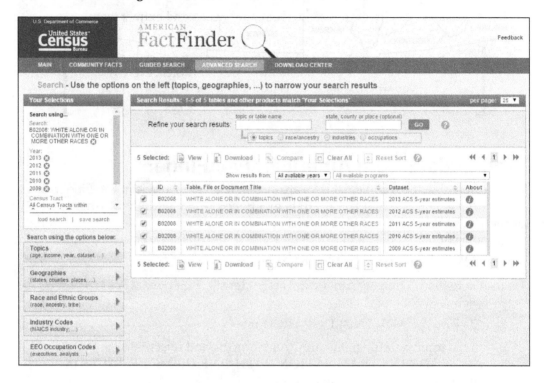

8. Now, remove **B02008: WHITE ALONE OR IN COMBINATION WITH ONE OR MORE RACES** from the search filter showing selections in the upper-left corner of the page.

9. Enter `total` into the **topic or table name** field, selecting **B01003: TOTAL POPULATION** from the suggested datasets, and then click on **GO**.

10. Select the five 2009 to 2013 total population 5-year estimates and then click on **GO**.

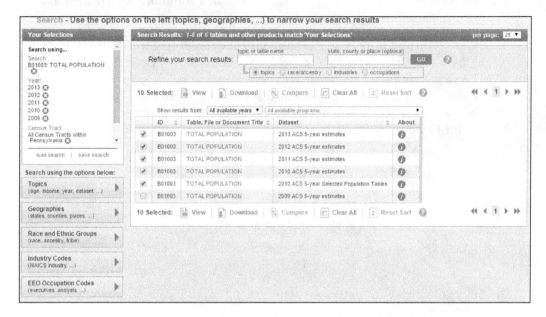

11. Click on **Download** to download these 10 datasets, as shown in the preceding screenshot.

12. Once you see the **Your file is complete** message, click on **DOWNLOAD** again to download the files. These will download as a `aff_download.zip` directory.

13. Move this directory to `c5/data/original` and then extract it.

Preparing and exporting the data

First, we will cover the steps for tabular data preparation and exporting, which are fairly similar to those we've done before. Next, we will cover the steps for preparing the boundary data, which will be more novel. We need to prepare this data based on the spatial relationships between layers, requiring the use of SQLite, since this cannot easily be done with the out-of-the-box or plugin functionality in QGIS.

The tabular data

Our tabular data is of the census tract white population. We only need to have the parseable latitude and longitude fields in this data for plotting later and, therefore, can leave it in this generic tabular format.

Combining it yearly

To combine this yearly data, we can join each table on a common GEOID field in QGIS. Perform the following steps:

1. Open QGIS and import all the boundary shapefiles (the tracts and state house boundaries) and data tables (all the census tract years downloaded). The single boundary shapefile will be in its extracted directory with the `.shp` extension. Data tables will be named something similar to `x_with_ann.csv`. You need to do this the same way you did earlier, which was through **Add Vector Layer** under the **Layer** menu. Here is a list of all the files to add:

    *   `tl_2014_42_tract.shp`
    *   `ACS_09_5YR B01003_with_ann.csv`
    *   `ACS_10_5YR B01003_with_ann.csv`
    *   `ACS_11_5YR B01003_with_ann.csv`
    *   `ACS_12_5YR B01003_with_ann.csv`
    *   `ACS_13_5YR B01003_with_ann.csv`

2. Select the tract boundaries shapefile, `tl_2014_42_tract`, from the **Layers** panel.

3. Navigate to **Layers | Properties**.

4. For each white population data table (ending in `x_B02008_with_ann`), perform the following steps:

    1. On the **Joins** tab, click on the green plus sign (**+**) to add a join.
    2. Select a data table as the **Join layer**.
    3. Select **GEO.id2** in the **Join field** tab.
    4. **Target field: GEOID**

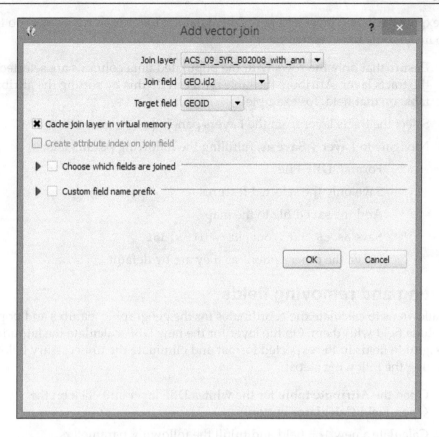

After joining all the tables, you will find many rows in the attribute table containing null values. If you sort them a few years later, you will find that we have the same number of rows populated for more recent years as we have in the Philadelphia tracts layer. However, the year 2009 (ACS_09_5YR B01003_with_ann.csv) has many rows that could not be populated due to the changes in the unique identifier used in the 2014 boundary data. For this reason, we will exclude the year 2009 from our analysis. You can remove the 2009 data table from the joined tables so that we don't have any issue with this later.

Now, export the joined layer as a new DBF database file, which we need to do to be able to make some final changes:

1. Ensure that only the rows with the populated data columns are selected in the tracts layer. Attribute the table (you can do this by sorting the attribute table on that field, for example).

2. Select the tracts layer from the **Layers** panel.

3. Navigate to **Layer | Save as**, fulfilling the following parameters:

   ° **Format: DBF File**

   ° Save only the selected features

   ° Add the saved file to the map

   ° **Save as**: `c5/data/output/whites.dbf`

   ° Leave the other options as they are by default

Updating and removing fields

QGIS allows us to calculate the coordinates for the geographic features and populate an attribute field with them. On the layer for the new DBF, calculate the latitude and longitude fields in the expected format and eliminate the unnecessary fields by performing the following steps:

1. Open the **Attribute table** for the whites DBF layer and click on the **Open Field Calculator** button.

2. Calculate a new `lon` field and fulfill the following parameters:

   ° **Output field name**: `lon`.

   ° **Output field type: Decimal number (real)**.

   ° **Output field width**: 10.

   ° **Precision**: 7.

   ° **Expression**: `"INTPLON"`. You can choose this from the **Fields** and **Values** sections in the tree under the **Functions** panel.

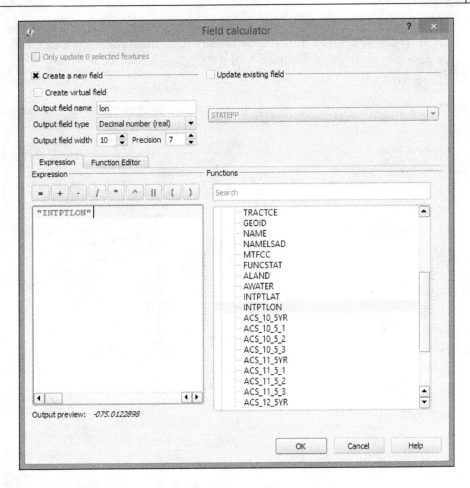

3. Repeat these steps with latitude, making a `lat` field from `INTPLAT`.

4. Create the following fields using the field calculator with the expression on the right:

    ° **Output field name**: name; **Output field type: Text; Output field width: 50; Expression**: NAMESLAD

    ° **Output field name**: Jan-11; **Output field type: Whole number (integer); Expression**: "ACS_11_5_2" - "ACS_10_5_2"

    ° **Output field name**: Jan-12; **Output field type: Whole number (integer); Expression**: "ACS_12_5_2" - "ACS_11_5_2"

    ° **Output field name**: Jan-13; **Output field type: Whole number (integer); Expression**: "ACS_13_5_2" - "ACS_12_5_2"

5. Remove all the old fields (except name, Jan-11, Jan-12, Jan-13, lat, and lon). This will remove all the unnecessary identification fields and those with a margin of error from the table.

6. Toggle the editing mode and save when prompted.

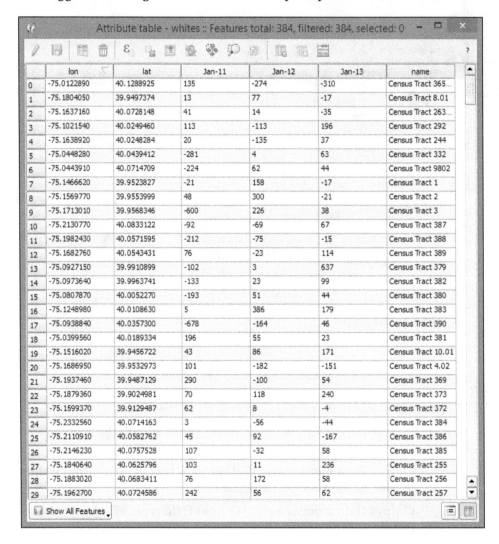

	lon	lat	Jan-11	Jan-12	Jan-13	name
0	-75.0122890	40.1288925	135	-274	-310	Census Tract 365...
1	-75.1804050	39.9497374	13	77	-17	Census Tract 8.01
2	-75.1637160	40.0728148	41	14	-35	Census Tract 263...
3	-75.1021540	40.0249460	113	-113	196	Census Tract 292
4	-75.1638920	40.0248284	20	-135	37	Census Tract 244
5	-75.0448280	40.0439412	-281	4	63	Census Tract 332
6	-75.0443910	40.0714709	-224	62	44	Census Tract 9802
7	-75.1466620	39.9523827	-21	158	-17	Census Tract 1
8	-75.1569770	39.9553999	48	300	-21	Census Tract 2
9	-75.1713010	39.9568346	-600	226	38	Census Tract 3
10	-75.2130770	40.0833122	-92	-69	67	Census Tract 387
11	-75.1982430	40.0571595	-212	-75	-15	Census Tract 388
12	-75.1682760	40.0543431	76	-23	114	Census Tract 389
13	-75.0927150	39.9910899	-102	3	637	Census Tract 379
14	-75.0973640	39.9963741	-133	23	99	Census Tract 382
15	-75.0807870	40.0052270	-193	51	44	Census Tract 380
16	-75.1248980	40.0108630	5	386	179	Census Tract 383
17	-75.0938840	40.0357300	-678	-164	46	Census Tract 390
18	-75.0399560	40.0189334	196	55	23	Census Tract 381
19	-75.1516020	39.9456722	43	86	171	Census Tract 10.01
20	-75.1686950	39.9532973	101	-182	-151	Census Tract 4.02
21	-75.1937460	39.9487129	290	-100	54	Census Tract 369
22	-75.1879360	39.9024981	70	118	240	Census Tract 373
23	-75.1599370	39.9129487	62	8	-4	Census Tract 372
24	-75.2332560	40.0714163	3	-56	-44	Census Tract 384
25	-75.2110910	40.0582762	45	92	-167	Census Tract 386
26	-75.2146230	40.0757528	107	-32	58	Census Tract 385
27	-75.1840640	40.0625796	103	11	236	Census Tract 255
28	-75.1883020	40.0683411	76	172	58	Census Tract 256
29	-75.1962700	40.0724586	242	56	62	Census Tract 257

Attribute table - whites :: Features total: 384, filtered: 384, selected: 0

Show All Features

Finally, export the modified table as a new CSV data table, from which we will create our map visualization. Perform the following steps:

1. Select the whites DBF layer from the **Layers** panel.

2. Navigate to **Layer | Save as** while fulfilling the following parameters:

    ° **Format: Comma Separated Value [CSV]**

    ° **Save as**: c5/data/output/whites.csv

    ° Leave the other options as they were by default

The boundary data

Although we have the boundary data for the census tracts, we are only interested in visualizing the State House Districts in our application. Our stakeholders are interested in visualizing change for these districts. However, as we do not have the population data by race for these boundary units, let alone by the yearly population, we need to leverage the spatial relationship between the State House Districts and the tracts to derive this information. This is a useful workflow whenever you have the data at a different level than the geographic unit you wish to visualize or query.

Calculating the average white population change in each census tract

Now, we will construct a field that contains the average yearly change in the white population between 2010 and 2013. Perform the following steps:

1. As mentioned previously, join the total population tables (ending in B01003_ with_ann) to the joined tract layer, t1_2014_42_tract, on the same **GEO. id2**, GEO fields from the new total population tables, and the tract layer respectively. Do not join the 2009 table, because we discovered that there were many null values in the join fields for the white-only version of this.

2. As before, select the 384 rows in the attribute table having the populated join columns from this table. Save only the selected rows, calling the saved shapefile dataset tract_change and adding this to the map.

3. Open the **Attribute table** and then open **Field Calculator**.

    ° Create a new field.

    ° **Output field name**: avg_change.

    ° **Output field type: Decimal number (real)**.

    ° **Output field width: 4, Precision: 2.**

° The following expression is the difference of each year from the previous year divided by the previous year to find the fractional change. This is then divided by three to find the average over three years and finally multiplied by 100 to find the percentage, as follows:

```
((("ACS_11_5_2" - "ACS_10_5_2")/ "ACS_10_5_2" )+
 (("ACS_12_5_2" - "ACS_11_5_2")/ "ACS_11_5_2" )+
 (("ACS_13_5_2" - "ACS_12_5_2")/ "ACS_12_5_2" ))/3 * 100
```

4. After this, click on **OK**.

The spatial join in SpatiaLite

Now that we have a value for the average change in white population by tract, let's attach this to the unit of interest, which are the State House Districts. We will do this by doing a spatial join, specifically by joining all the records that intersect our House District bounds to that House District. As more than one tract will intersect each State House District, we'll need to aggregate the attribute data from the intersected tracts to match with the single district that the tracts will be joined to.

We will use SpatiaLite for doing this. Similar to PostGIS for Postgres, SpatiaLite is the spatial extension for SQLite. It is file-based; rather than requiring a continuous server listening for connections, a database is stored on a file, and client programs directly connect to it. Also, SpatiaLite comes with QGIS out of the box, making it very easy to begin to use. As with PostGIS, SpatiaLite comes with a rich set of spatial relationship functions, making it a good choice when the existing plugins do not support the relationship we are trying to model.

 SpatiaLite is usually not chosen as a database for live websites because of some limitations related to multiuser transactions—which is why CartoDB uses Postgres as its backend database.

Creating a SpatiaLite database

To do this, perform the following steps:

1. Create a new SpatiaLite database.

2. Navigate to **Layer | Create Layer | New Spatialite Layer**.

3. Using the ellipses button (**...**), browse to and create a database at c5/data/output/district_join.sqlite.

4. After clicking on **Save**, you will be notified that a new database has been registered. You have now created a new SpatiaLite database. You can now close this dialog.

Importing layers to SpatiaLite

To import layers to SpatiaLite, you can perform the following steps:

1. Navigate to **Database | DB Manager | DB Manage**.

2. Click on the refresh button. The new database should now be visible under the SpatiaLite section of the tree.

3. Navigate to **Table | Import layer/file** (`tract_change` and `tl_2014_42_sldl`).

4. Click on **Update options**.

5. Select **Create single-part geometries instead of multi-part**.

6. Select **Create spatial index**.

7. Click on **OK** to finish importing the table to the database (you may need to hit the refresh button again for table to be indicated as imported).

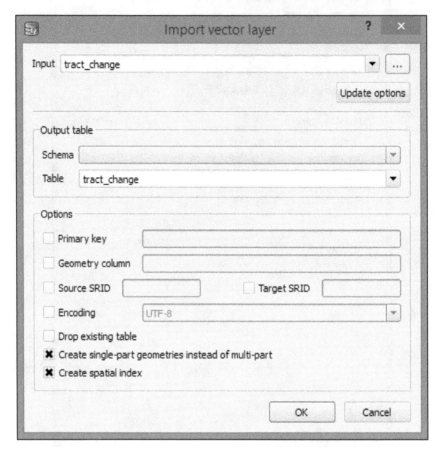

Now, repeat these steps with the House Districts layer (t1_2014_42_sldl), and deselect **Create single-part geometries instead of multi-part** as this seems to cause an error with this file, perhaps due to some part of a multi-part feature that would not be able to remain on its own under the SpatiaLite data requirements.

Querying and loading the SpatiaLite layer from the DB Manager

Next, we use the DB Manager to query the SpatiaLite database, adding the results to the QGIS layers panel.

We will use the `MBRIntersects` criteria here, which provides a performance advantage over a regular `Intersects` function as it only checks for the intersection of the extent (bounding box). In this example, we are dealing with a few features of limited complexity that are not done dynamically during a web request, so this shortcut does not provide a major advantage—we do this here so as to demonstrate its use for more complicated datasets.

1. If it isn't already open, open **DB Manager**.
2. Navigate to **Database | SQL window**.

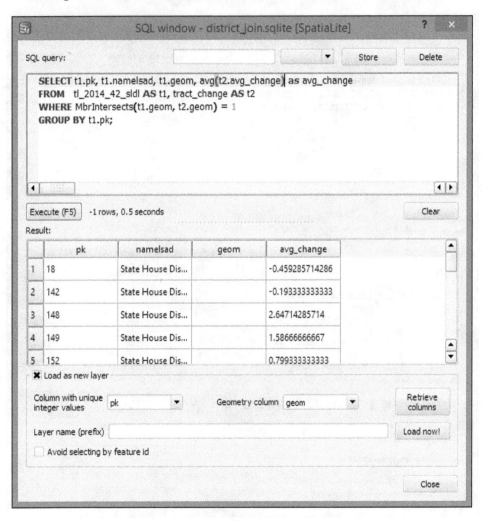

- ° Fill the respective input fields in the **SQL query** dialog:

° The following SQL query selects the fields from the `tract_change` and `tl_2014_42_sldl` (State Legislative District) tables, where they overlap. It also performs an aggregate (average) of the change by the State Legislative Districts overlying the census tract boundaries:

```
SELECT t1.pk, t1.namelsad, t1.geom, avg(t2.avg_change)*1.0
    as avg_change
FROM    tl_2014_42_sldl AS t1, tract_change AS t2
WHERE MbrIntersects(t1.geom, t2.geom) = 1
GROUP BY t1.pk;
```

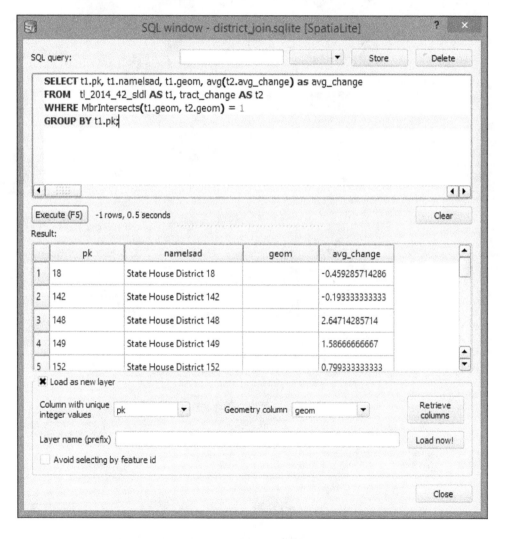

3. Then, click on **Load now!**.

4. You will be prompted to select a value for the **Column with unique integer values** field. For this, select **pk**.

5. You will also be prompted to select a value for the **Geometry column** field; for this, select **geom**.

The symbolized result of the spatial relationship join showing the average white population change over a 4-year period for the State House Districts' census tracts intersection will look something similar to the following image:

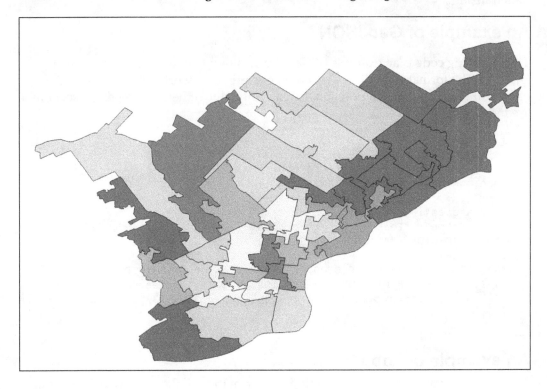

TopoJSON

Next, we will move on to preparing this data relationship for the Web and its spatiotemporal visualization.

TopoJSON is a variant of JSON, which uses the topological relationships between the geometric features to greatly reduce the size of the vector data and thereby improves the browser's rendering performance and reduces the risk of delay due to data transfers.

An example of GeoJSON

The following code is an example of GeoJSON, showing two of our State House Districts. The format is familiar — based on our previous work with JSON — with sets of coordinates that define a polygonal area grouped together. The repeated sections are marked by ellipses (…).

```
{
   "type": "FeatureCollection",
   "crs": { "type": "name", "properties": { "name":
"urn:ogc:def:crs:OGC:1.3:CRS84" } },

   "features": [
      { "type": "Feature", "properties": { "STATEFP": "42", "SLDLST":
"181", "GEOID": "42181", "NAMELSAD": "State House District
181", "LSAD": "LL", "LSY": "2014", "MTFCC": "G5220", "FUNCSTAT":
"N", "ALAND": 8587447.000000, "AWATER": 0.000000, "INTPTLAT":
"+39.9796799", "INTPTLON": "-075.1533540" }, "geometry": { "type":
"Polygon", "coordinates": [ [ [ -75.176782, 39.987337 ], [ … ] ]] } },
{…}] }
```

An example of TopoJSON

The following code is the corresponding representation of the same two State House Districts in TopoJSON, as discussed earlier.

Although this example uses the same coordinate system (WGS84/EPSG:4326) as that used before, it is expressed as simple pairs of abstract space coordinates. These are ultimately transformed into the WGS coordinate system using the scale and translate data in the transform section of the data.

By taking advantage of the shared topological relationships between geometric objects, the amount of data can be drastically reduced from 21K to 7K. That's a reduction of 2/3! You will see in the following code that each polygon is not clearly represented on its own but rather through these topological relationships. The repeated sections are marked by ellipses (...).

```
{"type":"Topology","objects":{"geojson":{"type":"GeometryCollection","
crs":{"type":"name","properties":{"name":"urn:ogc:def:crs:OGC:1.3:CRS
84"}},"geometries":[{"type":"Polygon","properties":{"STATEFP":"42","S
LDLST":"181","GEOID":"42181","NAMELSAD":"State House District 181","LS
AD":"LL","LSY":"2014","MTFCC":"G5220","FUNCSTAT":"N","ALAND":8587447
,"AWATER":0,"INTPTLAT":"+39.9796799","INTPTLON":"-075.1533540"},"arcs-
":[[0,1]]},{"type":"Polygon","properties":{"STATEFP":"42","SLDLST":
"197","GEOID":"42197","NAMELSAD":"State House District 197","LSAD":
"LL","LSY":"2014","MTFCC":"G5220","FUNCSTAT":"N","ALAND":8251026,"A
WATER":23964,"INTPTLAT":"+40.0054726","INTPTLON":"-075.1405813"},"-
arcs":[[-1,2]]}]}},"arcs":[[[903,5300],[…]]],"transform":{"sca
le":[0.000006659365934984,0.000006594359435944118],"transla
te":[-75.176782,39.961088]}}
```

Vector simplification

Similar to TopoJSON, Vector simplification removes the nodes in a line or polygon layer and will often greatly increase the browser's rendering performance while decreasing the file size and network transfer time.

As the vector shapes can have an infinite level of complexity, in theory, simplification methods can decrease the complexity by more than 99 percent while still preserving the perceivable shape of the geometry. In reality, the level of complexity at which perception becomes significantly affected will almost never be this high; however, it is common to have a very acceptable perception change at 90 percent complexity loss. The more sophisticated simplification methods have improved results.

Simplification methods

A number of simplification methods are commonly in use, each having strengths for particular data characteristics and outcomes. If one does not produce a good result, you can always try another. In addition to the method itself, you will also usually be asked to define a threshold parameter for an acceptable amount of complexity loss, as defined by the percentage of complexity lost, area, or some other measure.

- **Douglas-Peucker**: In this, the threshold affects the distance from which the original lines and edges of the polygons are allowed to change. This is useful when the nodes to be simplified are densely located but can lead to "pointy" simplifications.

- **Visvalingam / effective area**: In this, the point forming the triangle having the least area with two adjacent points is removed. The threshold affects how many times this criterion is applied. This has been described by Mike Bostock, the creator of TopoJSON among other things, as simplification with the criteria of the least perceptible change.

- **Visvalingam / weighted area**: In this, the point forming the vertex with the most acute angle is removed. The threshold affects how many times this criterion is applied. This method provides the "smoothest" result, as it specifically targets the "spikes".

The Visvalingam effective area method is the only method of simplification offered in the TopoJSON command-line tool that we will use. Mapshaper, a web-based tool that we will take a look at offers all these three methods but uses the weighted area method by default.

Other options

Other options affecting the data size are often offered alongside the simplification methods.

- **Repair intersections:** This option will repair the simplifications that cause the lines or polygon edges to intersect.

- **Prevent shape removal:** This option will prevent the simplification that would cause the removal of the (usually small) polygon shapes.

- **Quantization:** Quantization controls the precision of the coordinates. This is easier to think about when we are dealing with the coordinates in linear units. For obvious reasons, you may want to extend the precision to 1/5000 of a mile—getting the approximate foot precision. Also obviously, great precision comes at the cost of greater data size, so you should not overquantize where the application or source does not support such precision.

Simplifying for TopoJSON

Both Web and desktop TopoJSON conversion tools that we will use support these simplification options. That way, you can simplify a polygon at the same time as you reduce the data size through the topological relationship notation.

Simplifying for other outputs

If you wish to produce data other than for TopoJSON, you will need to find another way to do the simplification.

QGIS provides **Simplify Geometries** out of the box (navigate to **Vector** | **Geometry Tools** | **Simplify Geometries**), which does a Douglas-Peucker simplification. While it is the most popular method, it may not be the most effective one (see the following section for more).

The Simplify plugin offers a Visalvingam method in addition to Douglas-Peuker.

Converting to TopoJSON

There are a few options for writing TopoJSON. We will take a look at one for the desktop, which requires a software installation, and one via the web browser. As you might imagine, the desktop option will be more stable for doing anything in a customized way, which the web browser does not support, and is also more stable with the more complex feature sets. The web browser has the advantage of not requiring an install.

Web mapshaper

You can use the web-based **mapshaper** software from `http://www.mapshaper.org/` to convert from shapefile and other formats to TopoJSON and vice versa. Perform the following steps to convert the State Legislative Districts shapefile to TopoJSON:

1.   Open your browser and navigate to the mapshaper website.

2.   Optionally, select a different simplification method or try other options. This is not necessary for this example.

3.   Browse to select the Philadelphia State Legislative Districts shapefile (`c5/data/original/tl_2014_42_sldl/tl_2014_42_sldl.shp`) from your local computer or drag a file in. As the page indicates, Shapefile, GeoJSON, and TopoJSON are supported.

4.   Optionally, choose a simplification proportion from the slider bar (again, this is not needed for this example).

5.   Export as TopoJSON by clicking on the **Export** button at the top.

You will get the following screen in your browser:

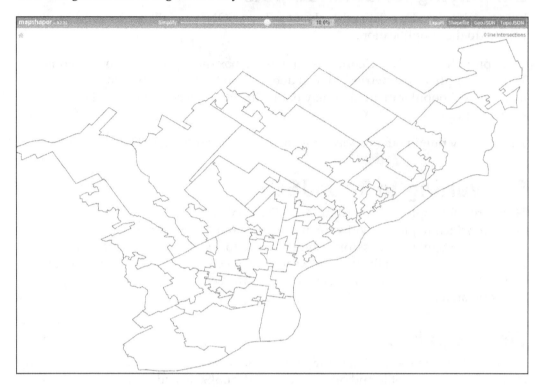

The command-line tool

The command-line tool is useful if you are working with a larger or more complicated dataset. The downside is that it requires that you install Node.js as it is a node package. For our purposes, Node.js is similar to Python. It is an interpreter environment for JavaScript, allowing the programs written in JavaScript to be run locally. In addition, it includes a package manager to install the needed dependencies. It also includes a web server—essentially running JavaScript as a server-side language.

Perform the following steps:

1. Install Node.js from `https://nodejs.org/`.
2. Open your OS command line (for example, on Windows, run cmd).
3. Input the following in the command line:
   ```
   >npm install -g topojson
   ```
4. Navigate to `cd c:\packt\c5\data\output` and input the following:
   ```
   >topojson -p -o house_district.json house_district.shp
   ```

You will now get the following output:

```
c:\packt\c5\data\temp>topojson -p -o house_district.json house_district.shp
bounds: -75.463053 39.848782 -74.869303 40.224734999999995 (spherical)
pre-quantization: 0.0660m (5.94e-7°) 0.0418m (3.76e-7°)
topology: 105 arcs, 16475 points
post-quantization: 6.60m (0.0000594°) 4.18m (0.0000376°)
prune: retained 105 / 105 arcs (100%)
```

Mapshaper also has a command-line tool, which we did not evaluate here.

The D3 data visualization library

D3 is a JavaScript library used for building the visualizations from the **Document Object Model (DOM)** present in all the modern web browsers.

What is D3?

In more detail, D3 manipulates the DOM into abstract vector visualization components, some of which have been further tailed to certain visualization types, such as maps. It provides us with the ability of parsing from some common data sources and binding, especially to the SVG and canvas elements that are designed to be manipulated for vector graphics.

Some fundamentals

There are a few basic aspects of D3 that are useful for you to understand before we begin. As D3 is not specifically built for geographic data, but rather for general data visualization, it tends to look at geographic data visualization more abstractly. Data must be parsed from its original format into a D3 object and rendered into the graphic space as an SVG or canvas element with a vector shape type. It must then be projected using relative mapping between the graphic space and a geographic coordinate system, scaled in relation to the graphic space and the geographic extent, and bound to a web object. This all must be done in relation to a D3 cursor of sorts, which handles the current scope that D3 is working in with keywords like "begin" and "end".

Parsing

We will be parsing through the d3.json and d3.csv methods. We use the callbacks of these methods to wrap the code that we want to be executed after the external data has been parsed into a JavaScript object.

Graphic elements, SVG, path, and Canvas

D3 makes heavy use of the two vector graphic elements in HTML5: SVG and Canvas. **Scaleable Vector Graphics (SVG)** is a mature technology for rendering vector graphics in the browser. It has seen some advancement in cross-browser support recently. Canvas is new to HTML5 and may offer better performance than SVG. Both, being DOM elements, are written directly as a subset of the larger HTML document rendered by the browser. Here, we will use SVG.

Projection

D3 is a bit unusual where geographic visualization libraries are concerned, in that it requires very little functionality specific to geographic data. The main geographic method provided is through the path element, projection, as D3 has its own concept of coordinate space, coordinates of the browser window and elements inside it.

Here is an example of projection. In the first line, we set the projection as Mercator. This allows us to center the map in familiar spherical latitude longitude coordinates. The scale property allows us to then zoom closer to the extent that we are interested in.

```
var projection = d3.geo.mercator()
  .center([-75.166667,40.03])
  .scale(60000);
```

Shape generator

You must configure a shape generator to bind to the d attribute of an SVG. This will tell the element how to draw the data that has been bound to it.

The main shape generator that we will use with the maps is path. Circle is also used in the following example, though its use is more complicated.

The following code creates a path shape generator, assigns it a projection, and stores it all in variable path:

```
var path = d3.geo.path()
  .projection(projection);
```

Scales

Scales allow the mapping of a domain of real data; say you have values of 1 through 100, in a range of possible values, and say you want everything down to numbers from 1 through 5. The most useful purpose of scales in mapping is to associate a range of values with a range of colors. The following code maps a set of values to a range of colors, mapping in-between values to intermediate colors:

```
var color = d3.scale.linear()
   .domain([-.5, 0, 2.66])
      .range(["#FFEBEB", "#FFFFEB", "#E6FFFF"]);
```

Binding

After a data object has been parsed into the DOM, it can be bound to a D3 object through its data or datum attribute.

Select, Select All, Enter, Return, Exit, Insert, and Append

In order to select the potentially existing elements, you will use the Select and Select All keywords. Then, based on whether you expect the elements to already be existent, you will use the Enter (if it is not yet existent), Return (if it is already existent), and Exit (if you wish to remove it) keywords to change the interaction with the element.

Here's an example of Select All, which uses the Enter keyword. The data from the `house_district` JSON, which was previously parsed, is loaded through the `d` attribute of the path element and assigned the path shape generator. In addition, a function is set on the `fill` attribute, which returns a color from the linear color scale:

```
map.selectAll("path")
   .data(topojson.feature(phila, phila.objects.house_district)
      .features)
   .enter()
      .append("path")
      .attr("vector-effect","non-scaling-stroke")
      .style("fill", function(d) { return color
         (d.properties.d_avg_change); })
      .attr("d", path);
```

Animated time series map

Through the following steps, we will produce an animated time series map with D3. We will start by moving our data to a filesystem path that we will use:

1. Move `whites.csv` to `c5/data/web/csv`.

2. Move `house_district.json` to `c5/data/web/json`.

The development environment

Start the Python HTTP server using the code from *Chapter 1, Exploring Places – from Concept to Interface*, (refer to the *Parsing the JSON data* section from *Chapter 7, Mapping for Enterprises and Communities*). This is necessary for this example, since the typical cross-site scripting protection on the browsers would block the loading of the JSON files from the local filesystem.

You will find the following files and directory structure under `c5/data/web`:

`./`	`index.html`
`./css/`	`main.css`
`./csv/`	`whites.csv` (you moved this here)
`./images/`	Various supporting images
`./js/`	`main.js`
`./json/`	`house_district.json` (you moved this here)
`./lib/`	• `d3.slider.js` • `d3.slider.css` • `d3.v3.min.js` • `topojson.v1.min.js`

Code

The following code, mostly JavaScript, will provide a time-based animation of our geographic objects through D3. This code is largely based on the one found at TIP Strategies' Geography of Jobs map found at `http://tipstrategies.com/geography-of-jobs/`. The main code file is at `c5/data/web/js/main.js`.

Note the reference to the CSV and TopoJSON files that we created earlier: `whites.csv` and `house_district.json`.

main.js

All of the following JavaScript code is in `./js/main.js`. All our customizations to this code will be done in this file:

```javascript
var width = 960,
  height = 600;

//sets up the transformation from map coordinates to DOM
  coordinates
var projection = d3.geo.mercator()
  .center([-75.166667,40.03])
  .scale(60000);

//the shape generator
var path = d3.geo.path()
  .projection(projection);

var svg = d3.select("#map-container").append("svg")
  .attr("width", width)
  .attr("height", height);

var g = svg.append("g");

g.append( "rect" )
  .attr("width",width)
  .attr("height",height)
  .attr("fill","white")
  .attr("opacity",0)
  .on("mouseover",function(){
    hoverData = null;
    if ( probe ) probe.style("display","none");
  })

var map = g.append("g")
  .attr("id","map");

var probe,
  hoverData;

var dateScale, sliderScale, slider;

var format = d3.format(",");

  var months = ["Jan"],
```

```
        months_full = ["January"],
        orderedColumns = [],
        currentFrame = 0,
        interval,
        frameLength = 1000,
        isPlaying = false;

var sliderMargin = 65;

function circleSize(d){
  return Math.sqrt( .02 * Math.abs(d) );
};

//color scale
var color = d3.scale.linear()
  .domain([-.5, 0, 2.66])
    .range(["#FFEBEB", "#FFFFEB", "#E6FFFF"]);

//parse house_district.json TopoJSON, reference color scale and
  other styles
d3.json("json/house_district.json", function(error, phila) {
  map.selectAll("path")
    .data(topojson.feature(phila, phila.objects.house_district)
      .features)
      .enter()
      .append("path")
      .attr("vector-effect","non-scaling-stroke")
      .attr("class","land")
      .style("fill", function(d) { return color(d.properties.
        d_avg_change); })
      .attr("d", path);

  //add a path element for district outlines
  map.append("path")
    .datum(topojson.mesh(phila, phila.objects.house_district,
      function(a, b) { return a !== b; }))
      .attr("class", "state-boundary")
      .attr("vector-effect","non-scaling-stroke")
      .attr("d", path);

  //probe is for popups
  probe = d3.select("#map-container").append("div")
    .attr("id","probe");

  d3.select("body")
```

```
  .append("div")
  .attr("id","loader")
  .style("top",d3.select("#play").node().offsetTop + "px")
  .style("height",d3.select("#date").node().offsetHeight +
    d3.select("#map-container").node().offsetHeight + "px");

//load and parse whites.csv
d3.csv("csv/whites.csv",function(data){
  var first = data[0];
  // get columns
  for ( var mug in first ){
    if ( mug != "name" && mug != "lat" && mug != "lon" ){
      orderedColumns.push(mug);
    }
  }
}

orderedColumns.sort( sortColumns );

// draw city points
for ( var i in data ){
  var projected = projection([ parseFloat(data[i].lon),
    parseFloat(data[i].lat) ])
  map.append("circle")
    .datum( data[i] )
    .attr("cx",projected[0])
    .attr("cy",projected[1])
    .attr("r",1)
    .attr("vector-effect","non-scaling-stroke")
    .on("mousemove",function(d){
      hoverData = d;
      setProbeContent(d);
      probe
      .style( {
        "display" : "block",
        "top" : (d3.event.pageY - 80) + "px",
        "left" : (d3.event.pageX + 10) + "px"
      })
    })
    .on("mouseout",function(){
      hoverData = null;
      probe.style("display","none");
    })
  }
```

```
createLegend();

dateScale = createDateScale(orderedColumns).range([0,3]);

createSlider();

d3.select("#play")
  .attr("title","Play animation")
  .on("click",function(){
    if ( !isPlaying ){
      isPlaying = true;
      d3.select(this).classed("pause",true).attr
        ("title","Pause animation");
      animate();
    } else {
      isPlaying = false;
      d3.select(this).classed("pause",false).attr
        ("title","Play animation");
      clearInterval( interval );
    }
  });

  drawMonth( orderedColumns[currentFrame] ); // initial map

  window.onresize = resize;
   resize();

  d3.select("#loader").remove();

})

});
```

Output

The finished product, which you can view by opening index.html in a web browser, is an animated set of points controlled by a timeline showing the change in the white population by the census tract. This data is displayed on top of the House Districts, colored from cool to hot by the change in the white population per year, and averaged over three periods of change (2010-11, 2011-12, and 2012-13). Our map application output, animated with a timeline, will look similar to this:

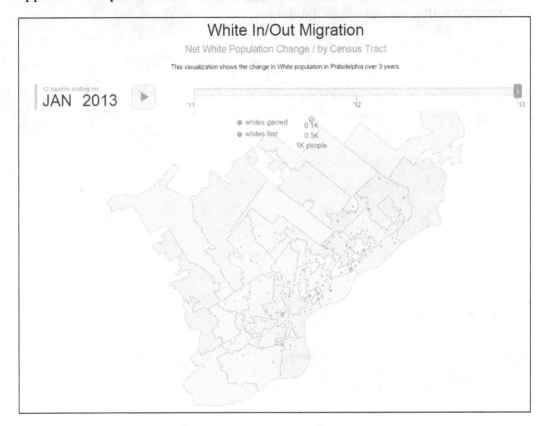

Summary

In this chapter, using an elections example, we covered spatial temporal data visualization and spatial relationship data integration. We also converted the data to TopoJSON, a format associated with D3, which greatly improves performance. We also created a spatial temporal animated web application through the D3 visualization library. In the next chapter, we will explore the interpolation to find the unknown values, the use of tiling, and the UTFGrid method to improve the performance with more complicated datasets.

6
Estimating Unknown Values

In this chapter, we will use interpolation methods to estimate the unknown values at one location based on the known values at other locations.

Interpolation is a technique to estimate unknown values entirely on their geographic relationship with known location values. As space can be measured with infinite precision, data measurement is always limited by the data collector's finite resources. Interpolation and other more sophisticated spatial estimation techniques are useful to estimate the values at the locations that have not been measured. In this chapter, you will learn how to interpolate the values in weather station data, which will be scored and used in a model of vulnerability to a particular agricultural condition: mildew. We've made the weather data a subset to provide a month in the year during which vulnerability is usually historically high. An end user could use this application to do a ground truthing of the model, which is, matching high or low predicted vulnerability with the presence or absence of mildew. If the model were to be extended historically or to near real time, the application could be used to see the trends in vulnerability over time or to indicate that a grower needs to take action to prevent mildew. The parameters, including precipitation, relative humidity, and temperature, have been selected for use in the real models that predict the vulnerability of fields and crops to mildew.

In this chapter, we will cover the following topics:

- Adding data from MySQL
- Using the NetCDF multidimensional data format
- Interpolating the unknown values for visualization and reporting
- Applying a simple algebraic risk model
- Python GDAL wrappers to filter and update through SQLite queries
- Interpolation
- Map algebra modeling

- Sampling a raster grid with a layer of gridded points
- Python CGI Hosting
- Testing and debugging during the CGI development
- The Python SpatiaLite/SQLite3 wrapper
- Generating an **OpenLayers3 (OL3)** map with the QGIS plugin
- Adding AJAX Interactivity to an OL3 map
- Dynamic response in the OL3 pixel popup

Importing the data

Often, the data to be used in a highly interactive, dynamic web application is stored in an existing enterprise database. Although these are not the usual spatial databases, they contain coordinate locations, which can be easily leveraged in a spatial application.

Connecting and importing from MySQL in QGIS

The following section is provided as an illustration only—database installation and setup are needlessly time consuming for a short demonstration of their use.

If you do wish to install and set up MySQL, you can download it from http://dev.mysql.com/downloads/. MySQL Community Server is freely available under the open source GPL license. You will want to install MySQL Workbench and MySQL Utilities, which are also available at this location, for interaction with your new MySQL Community Server instance. You can then restore the database used in this demonstration using the Data Import/Restore command with the provided backup file (c6/original/packt.sql) from MySQL Workbench.

To connect to and add data from your MySQL database to your QGIS project, you need to do the following (again, as this is for demonstration only, it does not require database installation and setup):

1. Navigate to **Layer | Add Layer | Add vector layer**.

    ○ **Source type: Database**

    ○ **Type: MySQL**, as shown in the following screenshot:

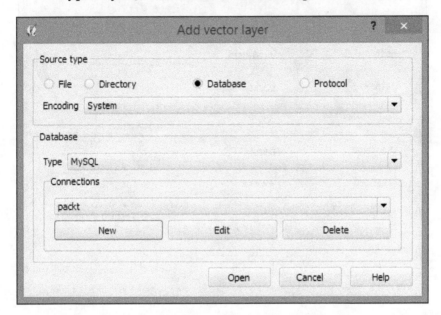

2. Once you've indicated that you wish to add a MySQL Database layer, you will have the option to create a new connection. In **Connections**, click on **New**. In the dialog that opens, enter the following parameters, which we would have initially set up when we created our MySQL Database and imported the .sql backup of the packt schema:

    ○ **Name:** packt

    ○ **Host:** localhost

    ○ **Database:** packt

    ○ **Port:** 3306

    ○ **Username:** packt

○ **Password**: packt, as shown in the following screenshot:

3. Click on **Test Connect**.

4. Click on **OK**.

5. Click on **Open**, and the **Select vector layers to add** dialog will appear.

6. From the **Select vector layers** dialog, click on **Select All**. This includes the following layers:

   ○ fields

   ○ precipitation

   ○ relative_humidity

   ○ temperature

7. Click on **OK**.

The layers (actually just the data tables) from the MySQL Database will now appear in the QGIS Layers panel of your project.

Converting to spatial format

The fields layer (table) is only one of the four tables we added to our project with latitude and longitude fields. We want this table to be recognized by QGIS as geospatial data and these coordinate pairs to be plotted in QGIS. Perform the following steps:

1. Export the fields layer as CSV by right–clicking on the layer under the **Layers** panel and then clicking on **Save as**.

2. In the **Save vector layer as…** dialog, perform the following steps:

   1. Click on **Browse** to choose a filesystem path to store the new .csv file. This file is included in the data under c6/data/output/fields.csv.

   2. For **GEOMETRY**, select **<Default>**.

   3. All the other default fields can remain as they are given.

   4. Click on **OK** to save the new CSV, as shown in the following screenshot:

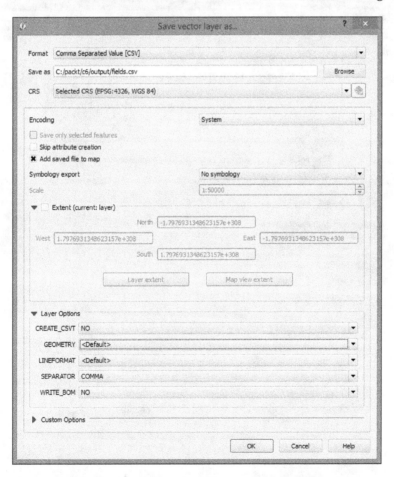

Now, to import the CSV with the coordinate fields that are recognized as geospatial data and to plot the locations, perform the following steps:

1. From the **Layer** menu, navigate to **Add Layer | Add Delimited Text Layer**.

2. In **Create a Layer from the Delimited Text File** dialog, perform the following steps:

   1. Click on the **Browse...** button to browse the location where you previously saved your fields.csv file (for example, c6/data/output/fields.csv).

   2. All the other parameters should be correctly populated by default. Take a look at the following image.

   3. Click on **OK** to create the new layer in your QGIS project.

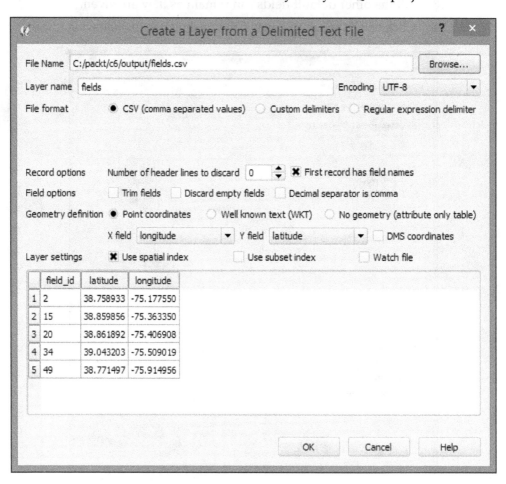

You will receive a notification that as no coordinate system was detected in this file, WGS 1984 was assigned. This is the correct coordinate system in our case, so no further intervention is necessary. After you dismiss this message, you will see the fields locations plotted on your map. If you don't, right–click on the new layer and select **Zoom to Layer**.

Note that this new layer is not reflected in a new file on the filesystem but is only stored with this QGIS project. This would be a good time to save your project.

Finally, join the other the other tables (`precipitation`, `relative_humidity`, and `temperature`) to the new plotted layer (fields) using the `field_id` field from each table one at a time. For a refresher on how to do this, refer to the *Table join* section of *Chapter 1*, *Exploring Places – from Concept to Interface*. To export each layer as separate shapefiles, right-click on each (`precipitation`, `relative_humidity`, and `temperature`), click on **Save as**, populate the path on which you want to save, and then save them.

The layer/table relations

The newer versions of QGIS support layer/table relations, which would allow us to model the one-to-many relationship between our locations, and an abstract measurement class that would include all the parameters. However, the use of table relationships is limited to a preliminary exploration of the relationships between layer objects and tables. The layer/table relationships are not recognized by any processing functions. Perform the following steps to explore the many-to-many layer/table relationships:

1. Add a relation by navigating to **Project | Project Properties | Relations**. The following image is what you will see once the relationships to the three tables are established:

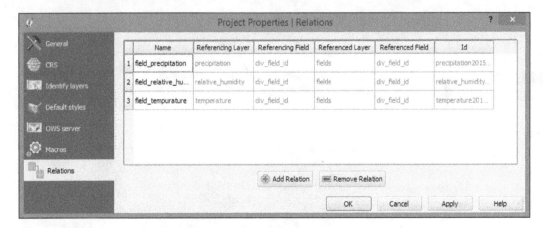

2. To add a relation, select a nonlayer table (for example, **precipitation**) in the **Referencing Layer (Child)** field and a location table (for example, **fields**) in the **Referenced Layer (Parent)** field. Use the common **Id** field (for example, field_id), which references the layer, to relate the tables. The name field can be filled arbitrarily, as shown in the following screenshot:

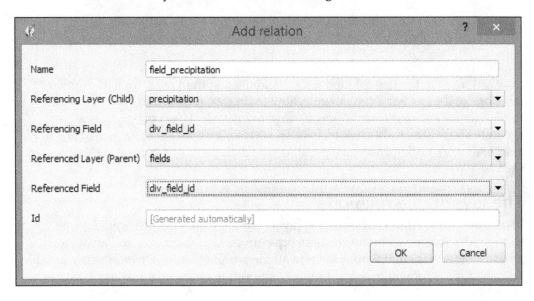

3. Now, to use the relation, click on a geographic object in the parent layer using the identify tool (you need to check **Auto open form** in the identify tool options panel). You'll see all the child entities (rows) connected to this object.

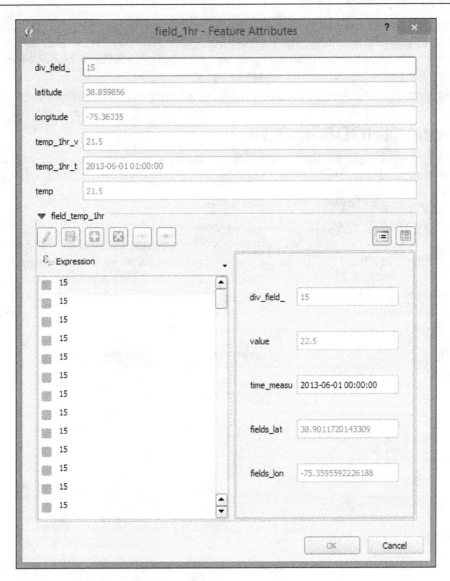

NetCDF

Network Common Data Form (NetCDF) is a standard—and powerful—format for environmental data, such as meteorological data. NetCDF's strong suit is holding multidimensional data. With its abstract concept of dimension, NetCDF can handle the dimensions of latitude, longitude, and time in the same way that it handles other often physical, continuous, and ordinal data scales, such as air pressure levels.

For this project, we used the monthly global gridded high-resolution station (land) data for air temperature and precipitation from 1901-2010, which the NetCDF University of Delaware maintains as part of a collaboration with NOAA. You can download further data from this source at `http://www.esrl.noaa.gov/psd/data/` `gridded/data.UDel_AirT_Precip.html`.

Viewing NetCDF in QGIS

While there is a plugin available, NetCDF can be viewed directly in QGIS, in GDAL via the command line, and in the QGIS Python Console. Perform the following steps:

1. Navigate to **Layer | Add Raster Layer**.

2. Browse to `c6/data/original/air.mon.mean.v301.nc` and add this layer.

3. Use the path **Raster | Miscellaneous > Information** to find the range of the values in a band. In the initial dialog, click on **OK** to go to the information dialog and then look for `air_valid_range`. You can see this information highlighted in the following image. Although QGIS's classifier will calculate the range for you, it is often thrown off by a numeric nodata value, which will typically skew the range to the lower end.

4. Enter the range information (-90 to 50) into the **Style** tab of the **Layer Properties** tab.

5. Click on **Invert** to show cool to hot colors from less to more, just as you would expect with temperature.

6. Click on **Classify** to create the new bins based on the number and color range. The following screenshot shows what an ideal selection of bins and colors would look like:

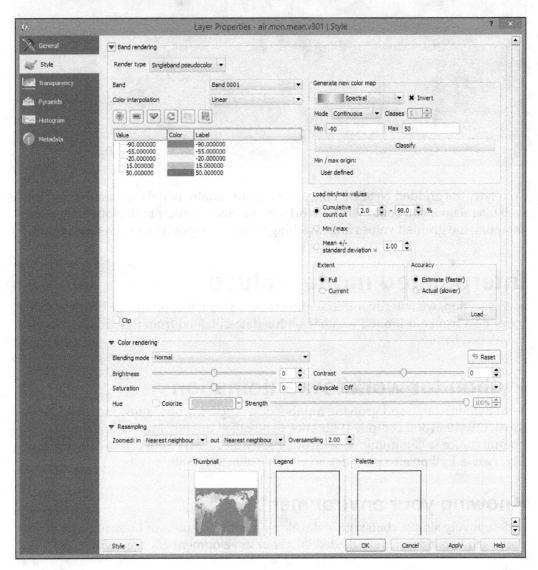

7.   Click on **OK**. The end result will look similar to the following image:

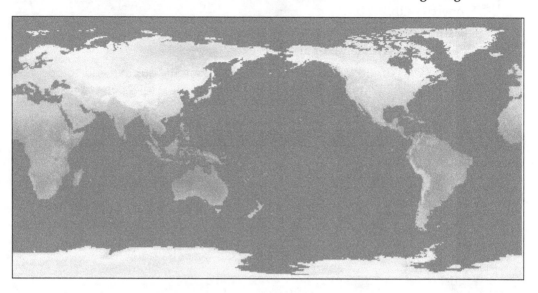

To render the gridded NetCDF data accessible to certain models, databases, and to web interaction, you could write a workflow program similar to the following after sampling the gridded values and attaching them to the points for each time period.

Interpolated model values

In this section, we will cover the creation of new statewide, point-based vulnerability index data from our limited weather station data obtained from the MySQL database mentioned before.

Python for workflow automation

With Python's large and growing number of wrappers, which allow independent (often C written and compiled) libraries to be called directly into Python, it is the natural choice to communicate with other software. Apart from direct API and library use, Python also provides access to system automation tasks.

Knowing your environment

A deceptively simple challenge in developing with Python is of knowing which paths and dependencies are loaded into your development environment.

To print your paths, type out the following in the QGIS Python Console (navigate to **Plugins | Python Console**) or the OSGeo4W Shell bundled with QGIS:

```
import os
try:
    user_paths = os.environ['PYTHONPATH'].split(os.pathsep)
except KeyError:
    user_paths = []
print user_paths
```

To list all the modules so that we know which are already available, type out the following:

```
import sys
sys.modules.keys()
```

Once we know which modules are available to Python, we can look up documentation on those modules and the programmable objects that they may expose.

Remember to view all the special characters (including whitespace) in whatever text editor or IDE you are using. Python is sensitive to indentation as it relates to code blocks! You can set your text editor to automatically write the tabs as a default number of spaces. For example, when I hit a tab to indent, I will get four spaces instead of a special tab character.

Generating the parameter grids for each time period

Now, we're going to move into nonevaluation code. You may want to take this time to quit QGIS, particularly if you've been working in the Python command pane. If I'm already working on the command pane, I like to quit using Python syntax with the following code:

```
quit()
```

After quitting, start QGIS up again. The Python Console can be found under **Plugins | Python Console**.

By running the next code snippet in Python, you will generate a command-line code, which we will run, in turn, to generate intermediate data for this web application.

What this code does

We will run a Python code to generate a more verbose script that will perform a lengthy workflow process.

- For each parameter (factor), it will loop through every day in the range of days. The range will effectively be limited to 06/10/15 through 06/30/15 as the model requires a 10-day retrospective period.

- We will run it via ogr2ogr—GDAL's powerful vector data transformation tool—and use the SQLite syntax, selecting the appropriate aggregate value (count, sum, and average) based on the relative period.

- It will translate each result by the threshold to scores for our calculation of vulnerability to mildew. In other words, using some (potentially arbitrary) breaks in the data, we will translate the real measurements to smaller integer scores related to our study.

- It will interpolate the scores as an integer grid.

Running a code in Python

Copy and paste the following lines into the Python interpreter. Press *Enter* if the code is pasted without execution. The code also assumes that data can be found in the locations hardcoded in the following (`C:/packt/c6/data/prep/ogr.sqlite`). You may need to move these files if they are not already in the given locations or change the code. You will also need to modify the following code according to your filesystem; Windows filesystem conventions are used in the following code:

```
# first variable to store commands
strCmds = 'del /F C:\packt\c6\data\prep\ogr.* \n'
# list of factors
factors = ['temperature','relative_humidity','precipitation']
# iterate through each factor, appending commands for each
for factor in factors:
  for i in range(10, 31):
    j = i - 5
    k = i - 9
    if factor == 'temperature':
      # commands use ogr2ogr executable from gdal project
      # you can run help on this from command line for more
      # information on syntax
      strOgr = 'ogr2ogr -f sqlite -sql "SELECT div_field_, GEOMETRY,
AVG(o_value) AS o_value FROM (SELECT div_field_, GEOMETRY, MAX(value)
AS o_value, date(time_measu) as date_f FROM {2} WHERE date_f BETWEEN
date(\'2013-06-{0:02d}\') AND date(\'2013-06-{1:02d}\') GROUP BY div_
field_, date(time_measu)) GROUP BY div_field_" -dialect sqlite -nln
ogr -dsco SPATIALITE=yes -lco SPATIAL_INDEX=yes -overwrite C:/packt/
c6/data/prep/ogr.sqlite C:/packt/c6/data/prep/temperature.shp \n'.
format(j,i,factor)
```

```
        strOgr += 'ogr2ogr -sql "UPDATE ogr SET o_value = 0 WHERE
o_value <=15.55" -dialect sqlite -update C:/packt/c6/data/prep/ogr.
sqlite C:/packt/c6/data/prep/ogr.sqlite \n'
        strOgr += 'ogr2ogr -sql "UPDATE ogr SET o_value = 3 WHERE
o_value > 25.55" -dialect sqlite -update C:/packt/c6/data/prep/ogr.
sqlite C:/packt/c6/data/prep/ogr.sqlite \n'
        strOgr += 'ogr2ogr -sql "UPDATE ogr SET o_value = 2 WHERE o_
value > 20.55 AND o_value <= 25.55" -dialect sqlite -update C:/packt/
c6/data/prep/ogr.sqlite C:/packt/c6/data/prep/ogr.sqlite \n'
        strOgr += 'ogr2ogr -sql "UPDATE ogr SET o_value = 1 WHERE o_
value > 15.55 AND o_value <= 20.55" -dialect sqlite -update C:/packt/
c6/data/prep/ogr.sqlite C:/packt/c6/data/prep/ogr.sqlite \n'
    elif factor == 'relative_humidity':
        strOgr = 'ogr2ogr -f sqlite -sql "SELECT GEOMETRY,
COUNT(value) AS o_value, date(time_measu) as date_f FROM relative_
humidity WHERE value > 96 AND date_f BETWEEN date(\'2013-06-{0:02d}\')
AND date(\'2013-06-{1:02d}\') GROUP BY div_field_" -dialect sqlite
-nln ogr -dsco SPATIALITE=yes -lco SPATIAL_INDEX=yes -overwrite C:/
packt/c6/data/prep/ogr.sqlite C:/packt/c6/data/prep/relative_humidity.
shp \n'.format(j,i)
        strOgr += 'ogr2ogr -sql "UPDATE ogr SET o_value = 0 WHERE o_
value <= 1" -dialect sqlite -update C:/packt/c6/data/prep/ogr.sqlite
C:/packt/c6/data/prep/ogr.sqlite \n'
        strOgr += 'ogr2ogr -sql "UPDATE ogr SET o_value = 3 WHERE o_
value > 40" -dialect sqlite -update C:/packt/c6/data/prep/ogr.sqlite
C:/packt/c6/data/prep/ogr.sqlite \n'
        strOgr += 'ogr2ogr -sql "UPDATE ogr SET o_value = 2 WHERE
o_value > 20 AND o_value <= 40" -dialect sqlite -update C:/packt/c6/
data/prep/ogr.sqlite C:/packt/c6/data/prep/ogr.sqlite \n'
        strOgr += 'ogr2ogr -sql "UPDATE ogr SET o_value = 1 WHERE
o_value > 10 AND o_value <= 20" -dialect sqlite -update C:/packt/c6/
data/prep/ogr.sqlite C:/packt/c6/data/prep/ogr.sqlite \n'
        strOgr += 'ogr2ogr -sql "UPDATE ogr SET o_value = 1 WHERE o_
value > 1 AND o_value <= 10" -dialect sqlite -update C:/packt/c6/data/
prep/ogr.sqlite C:/packt/c6/data/prep/ogr.sqlite \n'
    elif factor == 'precipitation':
        strOgr = 'ogr2ogr  -f sqlite -sql "SELECT GEOMETRY, SUM(value)
AS o_value, date(time_measu) as date_f FROM precipitation WHERE date_f
BETWEEN date(\'2013-06-{0:02d}\') AND date(\'2013-06-{1:02d}\') GROUP
BY div_field_" -dialect sqlite -nln ogr -dsco SPATIALITE=yes -lco
SPATIAL_INDEX=yes -overwrite C:/packt/c6/data/prep/ogr.sqlite C:/
packt/c6/data/prep/precipitation.shp \n'.format(k,i)
        strOgr += 'ogr2ogr -sql "UPDATE ogr SET o_value = 0 WHERE o_
value < 25.4" -dialect sqlite -update C:/packt/c6/data/prep/ogr.sqlite
C:/packt/c6/data/prep/ogr.sqlite \n'
        strOgr += 'ogr2ogr -sql "UPDATE ogr SET o_value = 3 WHERE o_
value > 76.2" -dialect sqlite -update C:/packt/c6/data/prep/ogr.sqlite
C:/packt/c6/data/prep/ogr.sqlite \n'
```

```
        strOgr += 'ogr2ogr -sql "UPDATE ogr SET o_value = 2 WHERE o_
value > 50.8 AND o_value <= 76.2" -dialect sqlite -update C:/packt/c6/
data/prep/ogr.sqlite C:/packt/c6/data/prep/ogr.sqlite \n'
        strOgr += 'ogr2ogr -sql "UPDATE ogr SET o_value = 1 WHERE o_
value > 30.48 AND o_value <= 50.8" -dialect sqlite -update C:/packt/
c6/data/prep/ogr.sqlite C:/packt/c6/data/prep/ogr.sqlite \n'
        strOgr += 'ogr2ogr -sql "UPDATE ogr SET o_value = 1 WHERE o_
value > 25.4 AND o_value <= 30.48" -dialect sqlite -update C:/packt/
c6/data/prep/ogr.sqlite C:/packt/c6/data/prep/ogr.sqlite \n'
        strGrid = 'gdal_grid -ot UInt16 -zfield o_value -l ogr -of
GTiff C:/packt/c6/data/prep/ogr.sqlite C:/packt/c6/data/prep/{0}
Inter{1}.tif'.format(factor,i)
        strCmds = strCmds + strOgr + '\n' + strGrid + '\n' + 'del /F
C:\packt\c6\data\prep\ogr.*' + '\n'

print strCmds
```

Running the printed commands in the Windows command console

Run the code output from the previous section by copying and pasting the result in the Windows command console. You can also find the output of the code to copy in `c6/data/output/generate_values.bat`.

The subprocess module

The subprocess module allows you to open up any executable on your system using the relevant command line syntax.

Although we could alternatively direct the code that we just produced through the subprocess module, it is simpler to do so directly on the command line in this case. With shorter, less sequential processes, you should definitely go ahead and use subprocess.

To use subprocess, just import it (ideally) in the beginning of your program and then use the `Popen` method to call your command line code. Execute the following code:

```
import subprocess
...
subprocess.Popen(strCmds)
```

Calculating the vulnerability index

GDAL_CALC evaluates an algebraic expression with gridded data as variables. In other words, you can use this GDAL utility to run map algebra or raster calculator type expressions. Here, we will use GDAL_CALC to produce our grid of the vulnerability index values based on the interpolated threshold scores.

Open a Python Console in QGIS (navigate to **Plugins** | **Python Console**) and copy/paste/run the following code. Again, you may wish to quit Python (using quit()) and restart QGIS/Python before running this code, which will produce the intermediate data for our application. This is used to control the unexpected variables and imported modules that are held back in the Python session.

After you've pasted the following lines into the Python interpreter, press *Enter* if it has not been executed. This code, like the previous one, produces a script that includes a range of numbers attached to filenames. It will run a map algebra expression through gdal_calc using the respective number in the range. Execute the following:

```
strCmd = ''

for i in range(10, 31):
    j = i - 5
    k = i - 9
    strOgr = 'gdal_calc --A C:/packt/c6/data/prep//temperatureInter{0}.
tif -B C:/packt/c6/data/prep/relative_humidityInter{0}.tif -C
C:/packt/c6/data/prep/precipitationInter{0}.tif --calc="A+B+C"
--type=UInt16 --outfile=C:/packt/c6/data/prep/calc{0}.tiff'.format(i)

    strCmd += strOgr + '\n'

print strCmd
```

Now, run the output from this code in the Windows command console. You can find the output code under c6/data/output/calculate_index.bat.

Creating regular points

As dynamic web map interfaces are not usually good at querying raster inputs, we will create an intermediate set of locations—points—to use for interaction with a user click event. The **Regular points** tool will create a set of points at a regular distance from each other. The end result is almost like a grid but made up of points. Perform the following steps:

1. Add c6/data/original/delaware_boundary.shp to your map project if you haven't already done so.

2. In **Vector**, navigate to **Research Tools | Regular points**.

3. Use **delaware_boundary** for the **Input Boundary Layer**.

4. Use a point spacing of .05 (in decimal degrees for now).

5. Save under c6/data/output/sample.shp.

The following image shows these parameters populated:

The output will look similar to this:

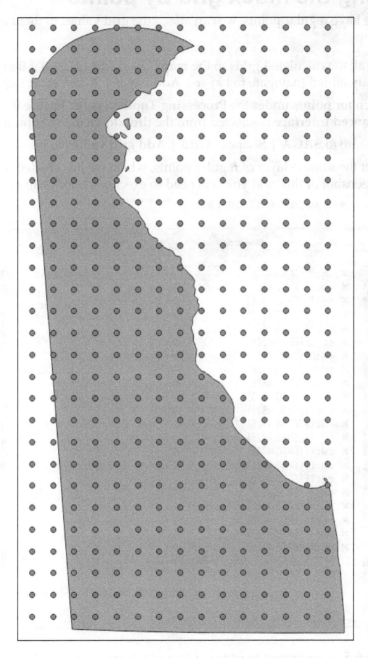

Sampling the index grid by points

Now that we have regular points, we can attach the grid values to them using the
following steps:

1. Add all the calculated grids to the map (`calc10` to `calc30`) if they were not
    already added (navigate to **Layer | Add Layer | Add Vector Layer**).

2. Search for **points** under the **Processing Toolbox** pane. Ensure that the
    Advanced Interface is selected from the dropdown at the bottom of the pane.

3. Navigate to **SAGA | Shapes - Grid | Add grid values to points**.

4. Select the **sample** layer of regular points, which we just created. Following is
    a screenshot of this, and you will need to execute the following code:

    ```
    Select all grids (calc10-calc30)
    ```

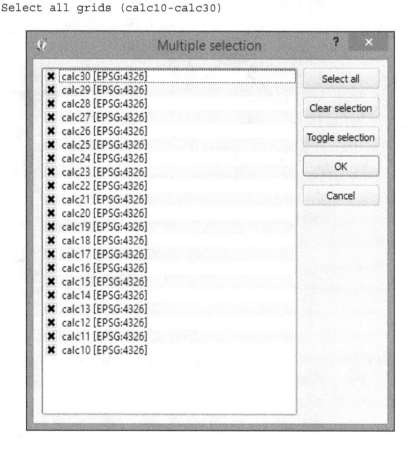

5. Save the output result to `c6/data/output/sample_data.shp`.

6. Click on **Run**, as shown in the following screenshot:

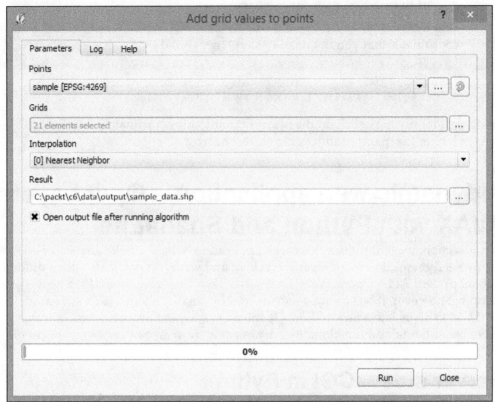

Create SQLite database and import

Next, create a SpatiaLite database at `c6/data/web/cgi-bin/c6.sqlite` (refer to the *Creating a SpatiaLite database* section of *Chapter 5, Demonstrating Change*) and import the `sample_data` shapefile using DB Manager.

DB Manager does not "see" SpatiaLite databases which were not created directly by the **Add Layer** command (as we've done so far; for example, in *Chapter 5, Demonstrating Change*), so it is best to do it this way rather than by saving it directly as a SpatiaLite database using the output dialog in the previous step.

Perform the following steps to test that our nearest neighbor result is correct:

1. Use the coordinate capture to get a test coordinate based on the points in the `sample_data` layer.

2. Create a SpatiaLite database using steps from *Chapter 5, Demonstrating Change* (navigate to **Layer | Create Layer**).

3. Open DB Manager (**Database | DB Manager**).

4. Import the `sample_data` layer/shapefile.

5. Run the following query in the DB Manager SQL window, substituting the coordinates that you obtained in step 1, separated by a space (for example, `75.28075 39.47785`):

```
SELECT pk, calc10, min(Distance(PointFromText('POINT (-
    75.28075 39.47785)'),geom)) FROM vulnerability
```

Using the identify tool, click on the nearest point to the coordinate you selected to check whether the query produces the correct nearest neighbor.

A dynamic web application – OpenLayers AJAX with Python and SpatiaLite

In this section, we will produce our web application, which, unlike any so far, involves a dynamic interaction between client and server. We will also use a different web map client API—OpenLayers. OpenLayers has long been a leader in web mapping; however, it has been overshadowed by smaller clients, such as Leaflet, as of late. With its latest incarnation, OpenLayers 3, OpenLayers has been slimmed down but still retains a functionality advantage in most areas over its newer peers.

Server side – CGI in Python

Common Gateway Interface (CGI) is perhaps the simplest way to run a server-side code for dynamic web use. This makes it great for doing proof of concept learning. The most typical use of CGI is in data processing and passing it onto the database from the web forms received through HTTP POST. The most common attack vector is the SQL injection. Going a step further, dynamic processing similar to CGI is often implemented through a minimal framework, such as Bottle, CherryPy, or Flask, to handle common tasks such as routing and sometimes templating, thus making for a more secure environment.

Don't forget that Python is sensitive to indents. Indents are always expressed as spaces with a uniform number per hierarchy level. For example, an `if` block may contain lines prefixed by four spaces. If the `if` block falls within a `for` loop, the same lines should be prefaced by 8 spaces.

Python CGI development

Next, we will start up a CGIHTTPServer hosting instance via a separate Windows console session. Then, we will work on the development of our server-side code— primarily through the QGIS Python Console.

Starting a CGI hosting

Starting a CGI session is simple— you just need to use the -m command line switch directly with Python, which loads the module as you might load a script. The following code starts CGIHTTPServer in port 8000. The current working directory will be served as the public web directory; in this case, this is C:\packt\c6\data\web.

In a new Windows console session, run the following:

```
cd C:\packt\c6\data\web
python -m CGIHTTPServer 8000
```

Testing the CGI hosting

Python (.py) CGI files can only run out of directories named either cgi or cgi-bin. This is a precaution to ensure that we intend the files in this directory to be publically executable.

To test this, create a file at c6/data/web/cgi-bin/simple_test.py with the following content:

 The first line is our shebang, which allows this file to be independently executable through the interpreter listed in the path on Unix systems. While this has no effect on Windows systems, where execution is handled through file associations, we will leave this here for interoperability.

```
#!/usr/bin/python

# Import the cgi, and system modules
import cgi, sys

# Required header that tells the browser how to render the HTML.
print "Content-Type: text/html\n\n"
print "Hello world"
```

You should now see the "Hello world" message when you visit http://localhost:8000/cgi-bin/simple_test.py on your browser. To debug on the client side, make sure you are using a browser-based web development view, plugin, or extension, such as Chrome's Developer Tools toolbar or Firefox's Firebug extension.

Debugging server-side code

Here are a few ways through which you can debug during Python CGI development:

- Use the Python Console in QGIS (navigate to **Plugins | Python Console**). You can run the Python code from your Python CGI Scripts here directly; however, this will fail for the scripts that rely on information passed through HTTP, but you can at least catch syntax errors, and you can populate it with the expected values to compare the result to what you're getting on a web browser. Sometimes, you'll want to quit QGIS to clear out the memory of the Python interpreter.

- A quicker way of doing this is to run your script in the command line with -d (verbose debugging). This will catch any issues that may not come up in interactive use, avoiding the variables that may have inadvertently been set in the same interactive session (substitute `index.py` with the name of your Python script). Run the following command from your command line shell:

```
python -d C:\packt\c6\data\web\cgi-bin\index.py
```

- If your Python CGI script is interacting with a database, you definitely need to test the queries through DB Manager SQL Window (or whichever database interface you prefer). It is often helpful to populate the queries with the expected values.

- Go to the following location in our web browser (substitute `index.py` with the name of your Python script):

```
localhost:8000/cgi-bin/index.py
```

Our Python server-side, database-driven code

Now, let's create a Python code to provide dynamic web access to our SQLite database.

PySpatiaLite

The PySpatiaLite module provides dbapi2 access to SpatiaLite databases. Dbapi2 is a standard library for interacting with databases from Python. This is very fortunate because if you use the dbapi2 connector from the sqlite3 module alone, any query using spatial types or functions will fail. The sqlite3 module was not built to support SpatiaLite.

Add the following to the preceding code. This will perform the following functions:

- Import the PySpatiaLite module and connect to our sqlite3/SpatiaLite database

- Use the connection as a context manager, which automatically commits the executed queries and rolls back in case of an error

- To test that the connection is working, use the SQLITE_VERSION() function in a SELECT query and print the result

The following code, appended to the preceding one, can be found at c6/data/web/cgi-bin/db_test.py. Make sure that the path in the code for the SQLite database file matches the actual location on your system.

```
# Import the pySpatiaLite module
from pySpatiaLite import dbapi2 as sqlite3
conn = sqlite3.connect('C:\packt\c6\data\web\cgi-bin\c6.sqlite')

# Use connection handler as context
with conn:
  c = conn.cursor()
  c.execute('SELECT SQLITE_VERSION()')

  data = c.fetchone()
  print data
  print 'SQLite version:{0}'.format(data[0])
```

You can view the following results in a web browser at http://localhost:8000/cgi-bin/db_test.py.

```
(u'3.7.17',) SQLite version:3.7.17
```

The first time that the data is printed, it is preceded by a u and wrapped in single quotes. This tells us that this is a unicode string (as our database uses unicode encoding). If we access element 0 in this string, we get a nonwrapped result.

The Python code for web access to SQLite through JSON

The following code performs the following functions:

- It connects to the database

- It issues a query to find the minimum distance location and field where the specified location and date are given

- It returns JSON with field value pairs based on the database result field names and values

You can find the code at c6/data/web/cgi-bin/get_json.py:

```python
#!/usr/bin/python

import cgi, cgitb, json, sys
from pySpatiaLite import dbapi2 as sqlite3

# Enables some debugging functionality
cgitb.enable()

# Creating row factory function so that we can get field names
# in dict returned from query
def dict_factory(cursor, row):
  d = {}
  for idx, col in enumerate(cursor.description):
    d[col[0]] = row[idx]
  return d

# Connect to DB and setup row_factory
conn = sqlite3.connect('C:\packt\c6\data\web\cgi-bin\c6.sqlite')
conn.row_factory = dict_factory

# Print json headers, so response type is recognized and correctly
decoded
print 'Content-Type: application/json\n\n'

# Use CGI FieldStorage object to retrieve data passed by HTTP GET
# Using numeric datatype casts to eliminate special characters
fs = cgi.FieldStorage()
longitude = float(fs.getfirst('longitude'))
latitude = float(fs.getfirst('latitude'))
day = int(fs.getfirst('day'))

# Use user selected location and days to find nearest location
(minimum distance)
# and correct date column
query = 'SELECT pk, calc{2} as index_value, min(Distance(PointFrom
Text(\'POINT ({0} {1})\'),geom)) as min_dist FROM vulnerability'.
format(longitude, latitude, day)

# Use connection as context manager, output first/only result row as
json
with conn:
  c = conn.cursor()
  c.execute(query)
  data = c.fetchone()
  print json.dumps(data)
```

You can test the preceding code by commenting out the portion that gets arguments from the HTTP request and setting these arbitrarily.

The full code is available at `c6/data/web/cgi-bin/json_test.py`:

```
# longitude = float(fs.getfirst('longitude'))
# latitude = float(fs.getfirst('latitude'))
# day = int(fs.getfirst('day'))
longitude = -75.28075
latitude = 39.47785
day = 15
```

If you browse to `http://localhost:8000/cgi-bin/json_test.py`, you'll see the literal JSON printed to the browser. You can also do the equivalent by browsing to the following URL, which includes these arguments: `http://localhost:8000/cgi-bin/get_json.py?longitude=-75.28075&latitude= 39.47785&day=15`.

```
{"pk": 260, "min_dist": 161.77454362713507, "index_value": 7}
```

The OpenLayers/jQuery client-side code

Now that our backend code and dependencies are all in place, it's time to move on to integrating this into our frontend interface.

Exporting the OpenLayers 3 map using QGIS

QGIS helps us get started on our project by allowing us to generate a working OpenLayers map with all the dependencies, basic HTML elements, and interaction event handler functionality. Of course, as with qgis2leaf, this can be extended to include the additional leveraging of the map project layers and interactivity elements.

The following steps will produce an OpenLayers 3 map that we will modify to produce our database-interactive map application:

1. Start a new QGIS map or remove all the layers from the current one.
2. Add `delaware_boundary.shp` to the map. Pan and zoom to the Delaware geographic boundary object if QGIS does not do so automatically.

3. Convert the `delaware_boundary` polygon layer to lines by navigating to **Vector | Geometry Tools | Polygons to lines**. Nonfilled polygons are not supported by **Export to OpenLayers**. The following image shows these inputs populated:

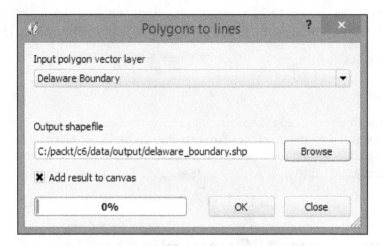

4. After you add the line boundaries, brighten them up and increase the size as well. Clicking on anything outside this boundary may not return a valid result. Rename the layer **Delaware Boundary** in the **Layers** panel.

5. Install the Export to OpenLayers 3 plugin if it isn't already installed.

6. Navigate to **Web | Export to OpenLayers | Create OpenLayers Map**.

7. In the **Export to OpenLayers 3** dialog, use the following parameter values (refer to the following image for clarification):

   ° Ensure that your stylized line boundary for Delaware (which we created in step 4) is checked and visible with no popup. Otherwise, this might obscure the interaction that we will create.

   ° Delete the unused fields. This is an important step, so uncheck this. Otherwise, you may see an error.

   ° If you've already zoomed and panned your canvas to the **Delaware Boundary** layer, you can ignore the **Extent** parameter. Otherwise, set the extent parameter to **Fit to Layers extent**.

   ° **Max zoom level**: 14.

   ° **Min zoom level**: 8.

   ° Select **Restrict to extent**.

○   Unselect **Use layer scale dependent visibility**.

○   **Base layer**: **MapQuest**, as shown in the following screenshot:

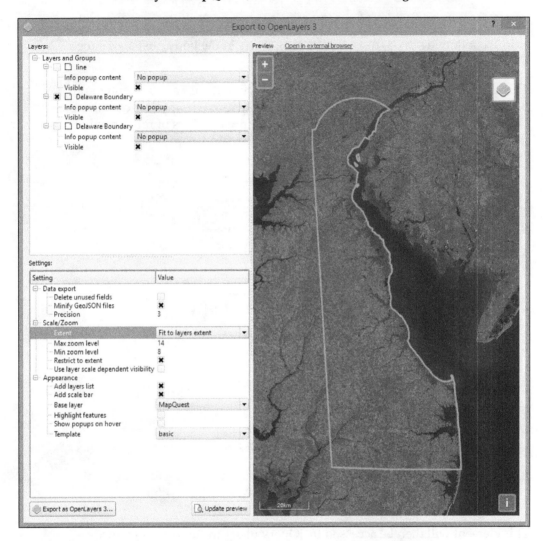

Modifying the exported OpenLayers 3 map application

Now that we have the base code and dependencies for our map application, we can move on to modifying the code so that it interacts with the backend, providing the desired information upon click interaction.

Remember that the backend script will respond to the selected date and location by finding the closest "regular point" and the calculated interpolated index for that date.

Adding an interactive HTML element

Add the following to `c6/data/web/index.html` in the body, just above the `div#id` element.

This is the HTML for the `select` element, which will pass a day. You would probably want to change the code here to scale with your application—this one is limited to days in a single month (and as it requires 10 days of retrospective data, it is limited to days from 6/10 to 6/30):

```
<select id="day">
  <option value="10">2013-06-10</option>
  <option value="11">2013-06-11</option>
  <option value="12">2013-06-12</option>
  <option value="13">2013-06-13</option>
  <option value="14">2013-06-14</option>
  <option value="15">2013-06-15</option>
  <option value="16">2013-06-16</option>
  <option value="17">2013-06-17</option>
  <option value="18">2013-06-18</option>
  <option value="19">2013-06-19</option>
  <option value="20">2013-06-20</option>
  <option value="21">2013-06-21</option>
  <option value="22">2013-06-22</option>
  <option value="23">2013-06-23</option>
  <option value="24">2013-06-24</option>
  <option value="25">2013-06-25</option>
  <option value="26">2013-06-26</option>
  <option value="27">2013-06-27</option>
  <option value="28">2013-06-28</option>
  <option value="29">2013-06-29</option>
  <option value="30">2013-06-30</option>
</select>
```

This element will then be accessed by jQuery using the `div#id` reference.

AJAX – the glue between frontend and backend

AJAX is a loose term applied specifically to an asynchronous interaction between client and server software using XML objects. This makes it possible to retrieve data from the server without the classic interaction of a submit button, which will take you to a page built on the result. Nowadays, AJAX is often used with JSON instead of XML to the same affect; it does not require a new page to be generated to catch the result from the server-side processing.

jQuery is a JavaScript library which provides many useful cross-browser utilities, particularly focusing on the DOM manipulation. One of the useful features that jQuery is known for is sending, receiving, and rendering results from AJAX calls. AJAX calls used to be possible from within OpenLayers; however, in OpenLayers 3, an external library is required. Fortunately for us, jQuery is included in the exported base OpenLayers 3 web application from QGIS.

Adding an AJAX call to the singleclick event handler

To add a jQuery AJAX call to our CGI script, add the following code to the "singleclick" event handler on SingleClick. This is our custom function that is triggered when a user clicks on the frontend map.

This AJAX call references the CGI script URL. The data object contains all the parameters that we wish to pass to the server. jQuery will take care of encoding the data object in a URL query string. Execute the following code:

```
jQuery.ajax({
  url: http://localhost:8000/cgi-bin/get_json.py,
  data: {"longitude": newCoord[0], "latitude": newCoord[1], "day":
    day}
})
Add a callback function to the jquery ajax call by inserting the
  following lines directly after it.
.done(function(response) {
popupText = 'Vulnerability Index (1=Least Vulnerable, 10=Most
  Vulnerable): ' + response.index_value;
```

Populating and triggering the popup from the callback function

Now, to get the script response to show in a popup after clicking, comment out the following lines:

```
/* var popupField;

  var currentFeature;
  var currentFeatureKeys;
  map.forEachFeatureAtPixel(pixel, function(feature, layer) {
    currentFeature = feature;
    currentFeatureKeys = currentFeature.getKeys();
    var field = popupLayers[layersList.indexOf(layer) - 1];
    if (field == NO_POPUP){
    }
    else if (field == ALL_FIELDS){
      for ( var i=0; i<currentFeatureKeys.length;i++) {
        if (currentFeatureKeys[i] != 'geometry') {
```

```
            popupField = currentFeatureKeys[i] + ': '+
               currentFeature.get(currentFeatureKeys[i]);
            popupText = popupText + popupField+'<br>';
         }
       }
    }
    else{
       var value = feature.get(field);
       if (value){
          popupText = field + ': '+ value;
       }
    }
}); */
```

Finally, copy and paste the portion that does the actual triggering of the popup in the
.done callback function. The .done callback is triggered when the AJAX call returns
a data response from the server (the data response is stored in the response object
variable). Execute the following code:

```
.done(function(response) {
  popupText = 'Vulnerability Index (1=Least Vulnerable, 10=Most
    Vulnerable): ' + response.index_value;
  if (popupText) {
    overlayPopup.setPosition(coord);
    content.innerHTML = popupText;
    container.style.display = 'block';
  } else {
    container.style.display = 'none';
   closer.blur();
  }
```

Testing the application

Now, the application should be complete. You will be able to view it in your browser
at http://localhost:8000.

You will want to test by picking a date from the **Select** menu and clicking on
different locations on the map. You will see something similar to the following
image, showing a susceptibility score for any location on the map within the study
extent (Delaware).

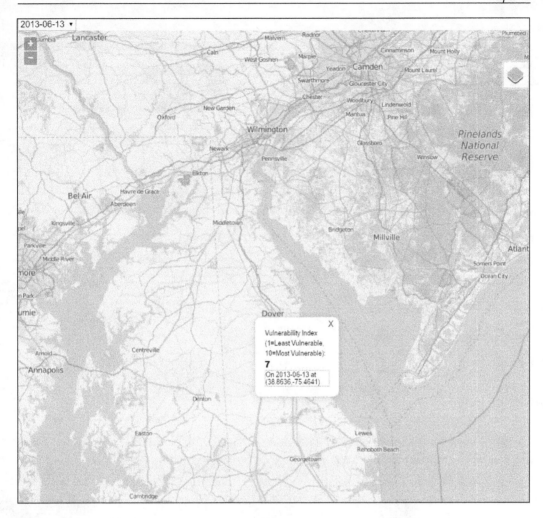

Summary

In this chapter, using an agricultural vulnerability modeling example, we covered interpolation and dynamic backend processing using spatial queries. The final application allows an end user to click on anything within a study area and see a score calculated from the interpolated point data and an algebraic model using a dynamic Python CGI code that queries a SQLite database. In the next chapter, we will continue to explore dynamic websites with an example that provides simple client editing capabilities and the use of the tiling and UTFGrid methods to improve performance with more complicated datasets.

7
Mapping for Enterprises and Communities

In this chapter, we will use a mix of web services to provide an editable collaborative data system.

While the visualization and data viewing capabilities that we've seen so far are a powerful means to reach an audience, we can tap into an audience—whether they are members of our organization, community stakeholders, or simply interested parties out on the web—to contribute improved geometric and attribute data for our geographic objects. In this chapter, you will learn to build a system of web services that provides these capabilities for a university community. As far as editable systems go, this is at the simpler end of things. Using a map server such as GeoServer, you could extend more extensive geometric editing capabilities based on a sophisticated user access management.

In this chapter, we will cover the following topics:

- Google Sheets for collaborative data management and services
- AJAX for web service processing
- OpenStreetMap for collaborative data contribution
- MBTiles and UTFGrid data formats
- Interactive data hosting through Mapbox
- Parsing and mapping JSON to an object
- Mixing web service data
- Setting up an Ubuntu virtual machine with Vagrant
- TileStream for local MBTiles hosting

Google Sheets for data management

Google Sheets provides us with virtually everything we need in a basic data management platform—it is web-based, easily editable through a spreadsheet interface, has fine-grained editing controls and API options, and is consumable through a simple JSON web service—at no cost, in most cases.

Creating a new Google document

To create a new Google document, you'll need to sign up for a Google account at `https://accounts.google.com`. Perform the following steps:

1. Create a new Google Sheets document at `https://docs.google.com/spreadsheets`.

2. Import data from an Excel file.

    1. Navigate to **File | Import**.

    2. Then, navigate to **Upload |** `c7/data/original/building_export.xlsx`.

Publishing Google Sheets on the Web

By default, Google Sheets will not be publicly viewable. In addition, no web service feed is exposed. To enable access to our data hosted by Google Sheets from our web application, we must publish the sheet. Perform the following steps:

1. Navigate to **File | Publish to the web**.

2. Copy and paste the URL (which appears after clicking on **Published**) to a location that you can refer to later (for example, in your favorite text editor).

3. Select the **Automatically republish when changes are made** checkbox if it is not already selected, as shown in the following screenshot:

 You'll need the section after **d/** (here, it starts with **1xAc8w**). This is the unique identifier referring to your sheet (or as it is sometimes known in documentation, the "key").

Previewing JSON

Now that we've published the sheet, our feed is exposed as JSON. We can view the JSON feed by substituting KEY with our spreadsheet unique identifier in a URL of the format `https://spreadsheets.google.com/feeds/list/KEY/1/public/basic?alt=json`. For example, it would look similar to the following URL:

`https://spreadsheets.google.com/feeds/list/1xAc8wpgLgTZpvZmZau20iO1dhA_31ojKSIBmlG6FMzQ/1/public/basic?alt=json`

This produces the following JSON response. For brevity, the response has been truncated after the first building object:

```
{"version":"1.0","encoding":"UTF-8","feed":{"xmlns":"http://www.
w3.org/2005/Atom","xmlns$openSearch":"http://a9.com/-/spec/opensearc
hrss/1.0/","xmlns$gsx":"http://schemas.google.com/spreadsheets/2006/
extended","id":{"$t":"https://spreadsheets.google.com/feeds/
list/19xiRHxZE4jOnVcMDXFx1pPyir4fXVGisWOc8guWTo2A/od6/public/
basic"},"updated":{"$t":"2012-04-06T13:55:10.774Z"},"category":[{"s
cheme":"http://schemas.google.com/spreadsheets/2006","term":"http://
schemas.google.com/spreadsheets/2006#list"}],"title":{"type":"tex
t","$t":"Sheet 1"},"link":[{"rel":"alternate","type":"application/
atom+xml","href":"https://docs.google.com/spreadsheets/d/19xiRHxZE4j
OnVcMDXFx1pPyir4fXVGisWOc8guWTo2A/pubhtml"},{"rel":"http://schemas.
google.com/g/2005#feed","type":"application/atom+xml","href":"https://
spreadsheets.google.com/feeds/list/19xiRHxZE4jOnVcMDXFx1pPyir4fX
VGisWOc8guWTo2A/od6/public/basic"},{"rel":"http://schemas.google.
com/g/2005#post","type":"application/atom+xml","href":"https://
spreadsheets.google.com/feeds/list/19xiRHxZE4jOnVcMDXFx1pPyir4fXVGi
sWOc8guWTo2A/od6/public/basic"},{"rel":"self","type":"application/
atom+xml","href":"https://spreadsheets.google.com/feeds/list/19xiRHxZE
4jOnVcMDXFx1pPyir4fXVGisWOc8guWTo2A/od6/public/basic?alt\u003djson"}],
"author":[{"name":{"$t":"Ben.Mearns"},"email":{"$t":"ben.mearns@gmail.
com"}}],"openSearch$totalResults":{"$t":"293"},"openSearch$startInde
x":{"$t":"1"},"entry":[{"id":{"$t":"https://spreadsheets.google.com/
feeds/list/19xiRHxZE4jOnVcMDXFx1pPyir4fXVGisWOc8guWTo2A/od6/public/
basic/cokwr"},"updated":{"$t":"2012-04-06T13:55:10.774Z"},"category"
:[{"scheme":"http://schemas.google.com/spreadsheets/2006","term":"h
ttp://schemas.google.com/spreadsheets/2006#list"}],"title":{"type":"
text","$t":"71219005"},"content":{"type":"text","$t":"udcode: NW92,
name: 102 Dallam Rd., type: Housing, address: 102 Dallam Road, _ciyn3:
19716, _ckd7g: 102 Dallam Road 19716, subcampus: WC"},"link":[{"rel"
:"self","type":"application/atom+xml","href":"https://spreadsheets.
google.com/feeds/list/19xiRHxZE4jOnVcMDXFx1pPyir4fXVGisWOc8guWTo2A/
od6/public/basic/cokwr"}]}, ...
]}}
```

Parsing the JSON data

To work with the JSON data from this web service, we will use jQuery's AJAX capabilities. Using the attributes of the JSON elements, we can take a look at how the data is rendered in HTML as a simple web page.

Starting up the server

Start up SimpleHTTPServer on port 8000 for c7/data/web on the Windows command line using the following commands:

```
cd c:\packt\c7\data\web
python -m SimpleHTTPServer 8000
```

Test parsing with jQuery

You can take a look at the following code (on the file system at c7/data/web/ gsheet.html) to test our ability to parse the JSON data:

```html
<html>
  <body>
    <div class="results"></div>
  </body>
  <script src="http://code.jquery.com/jquery-1.11.3.min.js">
    </script>
  <script>

    // ID of the Google Spreadsheet
    var spreadsheetID = "1xAc8wpgLgTZpvZmZau20iO1dhA_31ojKSIBmlG6FMzQ";

    // Make sure it is public or set to Anyone with link can view
    var url = "https://spreadsheets.google.com/feeds/list/" +
      spreadsheetID + "/1/public/values?alt=json";

    $.getJSON(url, function(data) {

      var entry = data.feed.entry;

      $(entry).each(function(){
        // Column names are name, type, etc.
        $('.results').prepend('<h2>'+this.gsx$name.$t+
          '</h2><p>'+this.gsx$type.$t+'</p>');
      });

    });

  </script>
```

You can preview this in a web browser at `http://localhost:8000/gsheet.html`. You'll see building names followed by types, as shown in the following image:

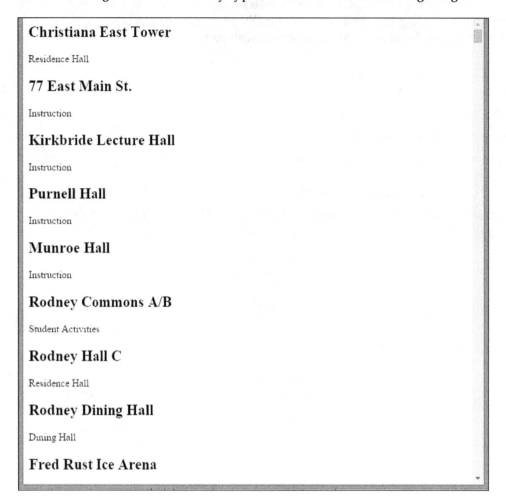

Rollout

Now let's take look at how we would operationalize this system for collaborative data editing.

Assigning permissions to additional users

In the sheet, click on the blue **Share** button in the upper-right corner. Alternatively, from Drive, select the file by clicking on it and then click on the icon that looks like a person with a plus sign on it. Ensure that anyone can find and view the document. Finally, add the address of the people you'd like to be able to edit the document and give them edit permissions, as shown in the following screenshot:

The editing workflow

Now that your collaborators have received an invitation to edit the sheet, they just need to sign in with their Google credentials and make a change to the sheet—the changes will be saved automatically. Of course, if they don't have any Google credentials, they'll need to create an account.

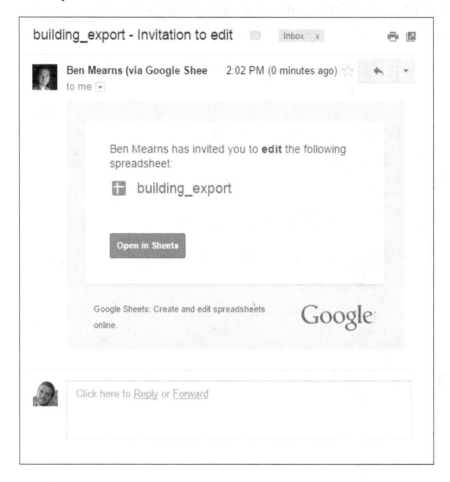

To go to the sheet, your collaborator will just need to click on **Open in Sheets**. The sheet should now also appear under their drive in **Shared with me**.

Here, you can see the type fields for **Christiana Hall**, **Kirkbride Lecture Hall**, and **Purnell Hall** after the changes are made:

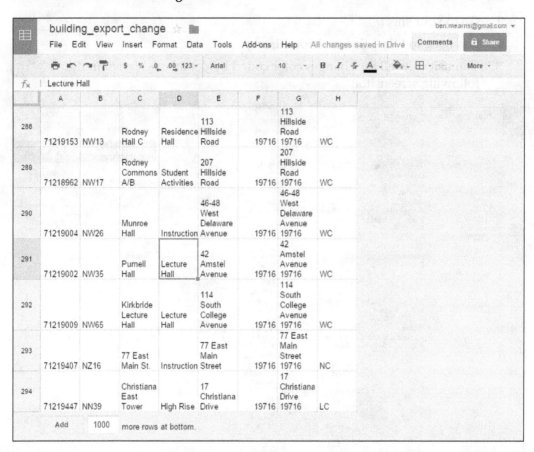

 If you don't require your collaborator to log in with Google, there is always the option of making your document publically editable—although, that comes with its own problems!

The publishing workflow

There is no need for an administrative intervention after the collaborators make changes. Data changed in sheets is automatically republished in the JSON feed, as we selected this option when we published the sheet. If you require more control over the publication of the collaborator edits, you may want to consider unselecting that option and setting up notifications of the changes. This way, you can republish after you've vetted the changes.

You can do a rollback of the changes as needed in the revision history. Perform the following steps:

1. Go to your sheet.

2. Navigate to **File | See revision history**.

3. You can view all the changes color coded by default, as shown in the following screenshot:

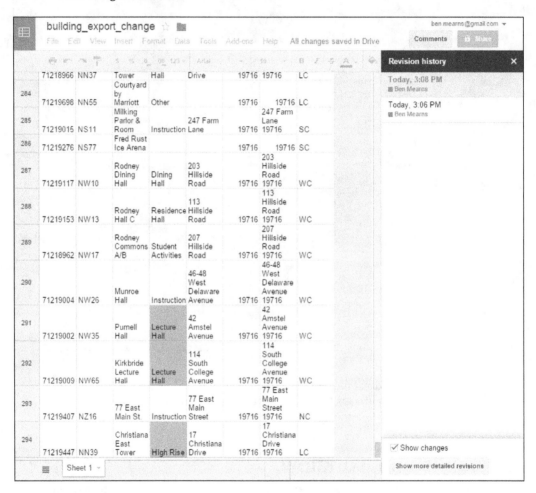

4. If you click on a particular change, you will have the option to restore the revision made to that point, as shown in the following screenshot:

Viewing the changes in your JSON feed

Go to `http://localhost:8000/gsheet.html` again to see how the changes to your sheet affected your JSON feed. Note in the following image the changes we made to the type fields for **Christiana East Tower**, **Kirkbride Lecture Hall**, and **Purnell Hall**:

Christiana East Tower

High Rise

77 East Main St.

Instruction

Kirkbride Lecture Hall

Lecture Hall

Purnell Hall

Lecture Hall

Munroe Hall

Instruction

Rodney Commons A/B

Student Activities

Rodney Hall C

Residence Hall

Rodney Dining Hall

Dining Hall

Fred Rust Ice Arena

In the final section of this chapter, we will also take a look at how we can preview this in the map interface.

The cartographic rendering of geospatial data – MBTiles and UTFGrid

At this point, you may be wondering, what about the maps? So far, we have not included any geospatial data or visualization. We will be offloading some of the effort in managing and providing geospatial data and services to OpenStreetMap—our favorite public open source geospatial data repository!

Why do we use OpenStreetMap?

- OSM already provides mirrored map services for quick reproduction in the basemaps

- OSM provides a very extensive and scalable schema for the kind of geographic features that you might find on a campus

- Various web, mobile, and desktop clients have already been written to interact with the OSM API

- OSM provides the databases and other infrastructure, so we don't have to

- OSM has a granular and reliable way to track changes, using the `osm_version` and `osm_user` fields, which complement the `osm_id` unique ID field

OpenStreetMap to SpatiaLite

To use the OSM data, we need to get it in a format that will be interoperable with other GIS software components. A quick and powerful solution is to store the OSM data in a SQLite SpatiaLite database instance, which, if you remember, is a single file with full spatial and SQL functionality.

To use QGIS to download and convert OSM to SQLite, perform the following steps:

1. Obtain the OSM data in the same way that we did in *Chapter 4, Finding the Best Way to Get There*. Use the OpenLayers plugin to zoom into Newark, DE (or use the extent, `39.7009, -75.7195, 39.6542, -75.7784`, clockwise from the top of the dialog in the next step):

    1. Navigate to **Vector | OpenStreetMap | Download Data** to download the OSM data for this extent.

2. Next, export the XML data in the `.osm` file to a topological SQLite database. This could potentially be used for routing; although, we will not be doing so here.

    1. Navigate to **Vector | OpenStreetMap | Import Topology from XML**.

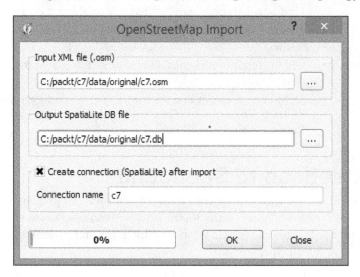

3. Next, export the topological data to normal geospatial data—polygons in this case.

    1. Navigate to **Vector | OpenStreetMap | Export topology to SpatiaLite**.

    2. **Export type: Polygons (closed ways)**.

    3. Click on **Load from DB** to populate the list of fields in the data. Select the fields **amenity**, **building**, **name**, and **leisure**, as shown in the following screenshot, as fit allowed:

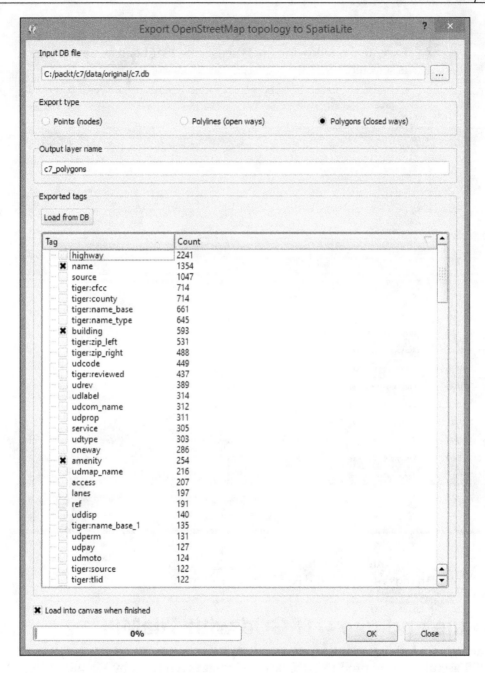

4. Use DB Manager to display the university buildings.

    1. Navigate to **Database | DB Manager | DB Manager**.

2. Highlight the c7 SQLite database.

3. Execute the following query, ensuring that **Load as new layer** is selected:

```
SELECT * FROM c7_polygons WHERE building = 'yes' and
    amenity = 'university'
```

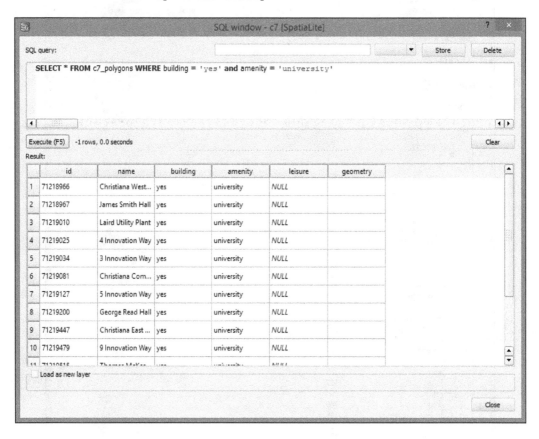

5. Export the query layer to `c7/data/original/delaware-latest-3875/buildings.shp` with the EPSG:3857 projection.

To tile and use UTFGrid with TileMill

Although TileMill is no longer under active production by its creator Mapbox, it is still useful for us to produce MBTiles tiled images rendered by Mapnik using CartoCSS and a UTFGrid interaction layer.

Preparing a basemap from OSM

TileMill requires that all the data be rendered and tiled together and, therefore, only supports vector data input, including JSON, shapefile, SpatiaLite, and PostGIS.

In the following steps, we will render a cartographically pleasing map as a .mbtiles (single-file-based) tile cache:

1. Install and open TileMill.

2. Download the Delaware data from the North America section of the Geofabrik OSM extracts site (http://download.geofabrik.de/north-america.html) as a shapefile. Alternatively, you can directly download it from http://download.geofabrik.de/north-america/us/delaware-latest.shp.zip. Ensure that you expand and copy the zip archive to your project directory after you've downloaded it.

3. Reproject all the data from EPSG:4326 to :3875. If you remember, QGIS can do this in batch as with other **Processing Toolbox** algorithms, as you learned in *Chapter 2*, *Identifying the Best Places*, making this process a bit quicker.

    Output all the layers to c7/data/original/delaware-latest-3875.

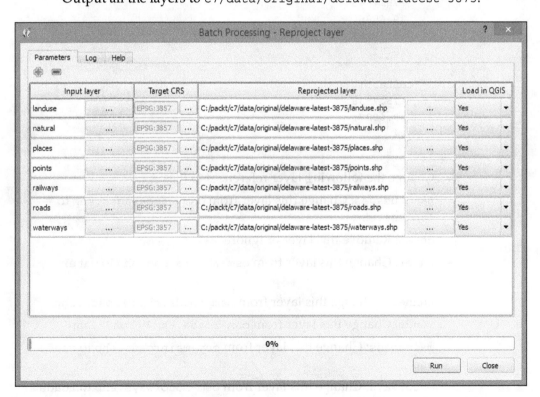

4. Copy the DC example to a new project.
   - ° You will find it in `C:\Program Files (x86)\TileMill-v0.10.1\tilemill\examples\open-streets-dc`
   - ° Copy it to `C:\Users\[YOURUSERNAME]\Documents\MapBox\project\c7`

5. Delete all the files from the `layers` directory.

6. Copy and extract all the shapefiles from `c7/data/original/delaware-latest-3875` into the `layers` directory in the `project` directory of `c7`, which can be found at `C:\Users\[YOURUSERNAME]\Documents\MapBox\project\c7\layers`.

7. Edit the `project.mml` file.
   1. Change all the instances of the `open-streets-dc` string to `c7`.
   2. Change the single instance of `Open Streets, DC` to `c7`.
   3. Substitute the following `bounds` and `center`:
      ```
      "bounds": [
        -75.7845,
        39.6586,
        -75.7187,
        39.71
      ],
      "center": [
        -75.7538,
        39.6827,
        14
      ],
      ```
   4. Change the following layer references to files:

      `Land usages`: Change this layer from `osm-landusages.shp` to `landuse.shp`

      `ocean`: Remove this layer or ignore

      `water`: Change this layer from `osm-waterareas.shp` to `waterways.shp`

      `tunnels`: Change this layer from `osm-roads.shp` to `roads.shp`

      `roads`: Change this layer from `osm-roads.shp` to `roads.shp`

      `mainroads`: Change this layer from `osm-mainroads.shp` to `roads.shp`

      `motorways`: Change this layer from `osm-motorways.shp` to `roads.shp` `bridges`: Change this layer from `osm-roads.shp` to `roads.shp`

`places`: Change this layer from `osm-places.shp` to `places.shp`

`road-label`: Change this layer from `osm-roads.shp` to `roads.shp`

8. Open TileMill and select the `c7` project from the **Projects** dialog, as shown in the following screenshot:

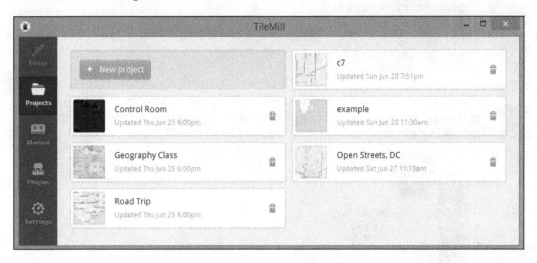

Preparing the operational layer in TileMill

1. Open the **Layers** panel from the bottommost button in the bottom-left corner. Refer to the next image.

2. Click on **+ Add layer**.

3. Populate the parameters with the following values:

    ° **ID**: `buildings`.

    ° **Datasource**: `c7/data/original/delaware-latest-3875/` `buildings.shp`.

    ° Click on Save & Style. You can return to this dialog later by clicking on the **Editor** button (pencil icon) in the **Layers** panel, by the `#c7` layer, as shown in the next image.

4. If you don't yet see your layer, ensure that you have some style defined in the tab on the right that will be applied to the layer (this should be populated by default with a minimal style). Then, click on **Save** in the top-right corner.

5. Use the CartoCSS syntax to change the style in `style.mss`. TileMill provides a color picker, which we can access by clicking on a swatch color at the bottom of the CartoCSS/style pane. After changing a color, you can view the hex code down there. Just pick a color, place the hex code in your CartoCSS, and save it. For example, consider the following code:

```
#buildings {
  line-color:#eb8f65;
  line-width:0.5;
  polygon-opacity:1;
  polygon-fill:#fdedc9;
}
```

6. Click on **Save** (with the pencil icon) in the upper-right corner of the main screen (above the CartoCSS input) to view the changes, as shown in the following screenshot:

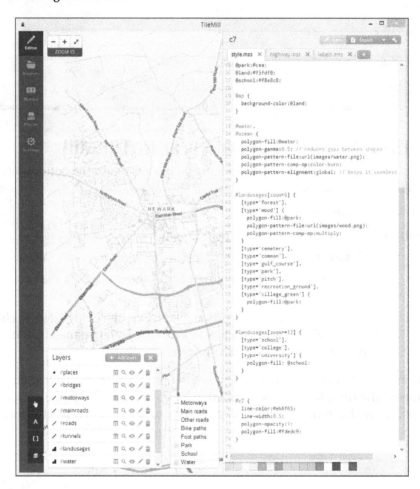

7. Go to the **Templates** tab by clicking on the topmost button in the lower-left corner and change the **Teaser** and **Full** interaction types to use {{{id}}} from buildings, as shown in the following screenshot:

Exporting MBTiles

MBTiles is a format developed by Mapbox to store geographic information. There are two compelling aspects of this format, besides interaction with a small but impressive suite of software and services developed by Mapbox: firstly, MBTiles stores a whole tile store in a single file, which is easy to transfer and maintain and secondly, UTFGrid, which is the use of UTF characters for highly performant data interaction, is enabled by this format.

Uploading to Mapbox

Perform the following steps:

1. Create an account on mapbox.com.

2. Access the **Export** dialog from the **Export** button in the upper-right corner. Select **Upload** from this menu.

3. Sign in to your Mapbox account by clicking on the button at the top of the dialog.
4. Press *Shift*, click on it, and drag to define an extent in the map.
5. Zoom to one level above your intended minimum zoom to preview the extent.
6. Fill in the descriptive information in the export dialog.
    ◦ **Name**: c7
    ◦ **Zoom**: 11 to 16
7. Click on the map to establish a **Center** coordinate.
8. Select **Save settings to project**.
9. Upload, as shown in the following screenshot:

The MBTiles file

The steps for exporting directly to an MBTiles file are similar to the previous procedure. This format can be uploaded to mapbox.com or served with software that supports the format, such as TileStream. Of course, no sign-on is needed.

Interacting with Mapbox services

In the last part of the previous section, we uploaded our rendered map to Mapbox in the MBTiles format.

To view the HTML page that Mapbox generates for our MBTiles, navigate to **Export | View Exports**. You'll find the upload listed there. Click on **View** to open it on a web browser.

For example, consider the following URL: `http://a.tiles.mapbox.com/v3/bmearns.c7/page.html#11/39.8715/-75.7514`.

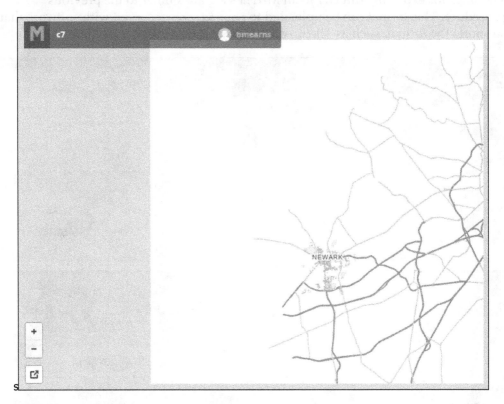

You may also want to preview the TileJSON web service connected to your data. You can do so by adding `.json` after your web map ID (`bmearns.c7`); for example, this is done in `http://a.tiles.mapbox.com/v3/bmearns.c7.json`, which executes the following:

```
{"attribution":"","bounds":[-75.8537,39.7943,-
75.6519,39.9376],"center":[-75.7514,39.8715,11],"description":"p
ackt, qgis, c7","download":"http://a.tiles.mapbox.com/v3/bmearns.
c7.mbtiles","embed":"http://a.tiles.mapbox.com/v3/bmearns.c7.html","f
ilesize":15349760,"format":"png","grids":["http://a.tiles.mapbox.com/
v3/bmearns.c7/{z}/{x}/{y}.grid.json","http://b.tiles.mapbox.com/v3/
bmearns.c7/{z}/{x}/{y}.grid.json"],"id":"bmearns.c7","legend":"","ma
xzoom":16,"minzoom":11,"modified":1435689144009,"name":"c7","private
":true,"scheme":"xyz","template":"{{#__location__}}{{/__location__}}
{{#__teaser__}}{{{id}}}{{/__teaser__}}{{#__full__}}{{{id}}}{{/__full__
}}","tilejson":"2.0.0","tiles":["http://a.tiles.mapbox.com/v3/bmearns.
c7/{z}/{x}/{y}.png","http://b.tiles.mapbox.com/v3/bmearns.c7/{z}/{x}/
{y}.png"],"version":"1.0.0","webpage":"http://a.tiles.mapbox.com/v3/
bmearns.c7/page.html"}
```

Connecting your local app with a hosted service

Now that we can see our tile server via Mapbox-generated HTML and JSON, let's take a look at how we can connect this with a local HTML that we can customize.

The API token

First, you'll need to obtain the API token from Mapbox. The token identifies your web application with Mapbox and enables the use of the web service you've created. As you will be adding this to the frontend code of your application, it will be publically known and open to abuse. Given Mapbox's monthly view usage limitations, you may want to consider a regular schedule for the token rotation (which includes creation, code modification, and deletion). This is also a good reason to consider hosting your service locally with something similar to TileStream, which is covered in the following *Going further – local MBTiles hosting with TileStream* section.

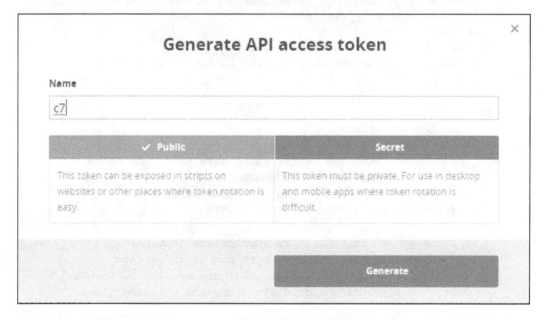

Mapbox.js

Mapbox.js is the mapping library developed by Mapbox to interact with its services. As Leaflet is at its core, the code will look familiar. We'll look at the modifications to the Creating a popup from UTFGrid data sample app code that you can get at https://www.mapbox.com/mapbox.js/example/v1.0.0/utfgrid-data-popup/.

Simple UTFGrid modification

In the following example, we will modify just the portions of code that directly reference the example data. Of course, we would want to change these portions to reference our data instead:

```html
<!DOCTYPE html>
<html>
  <head>
    <meta charset=utf-8 />
    <title>Creating a popup from UTFGrid data</title>
    <meta name='viewport' content='initial-scale=1,maximum-
      scale=1,user-scalable=no' />
    <script src='https://api.tiles.mapbox.com/mapbox.js/v2.2.1/
      mapbox.js'></script>
    <link href='https://api.tiles.mapbox.com/mapbox.js/v2.2.1/
      mapbox.css' rel='stylesheet' />
    <style>
      body { margin:0; padding:0; }
      #map { position:absolute; top:0; bottom:0; width:100%; }
    </style>
  </head>
<body>
    <div id='map'></div>
    <script>
      // changed token and center coordinate pair/zoom below
      L.mapbox.accessToken = 'YOURMAPBOXTOKEN';
      var map = L.mapbox.map('map', 'mapbox.streets')
        .setView([39.87240,-75.75367], 15);

      // change variable names as appropriate (optional)
      // changed mapbox id to refer to our layer/service
      var c7Tiles = L.mapbox.tileLayer('bmearns.c7').addTo(map);
      var c7Grid = L.mapbox.gridLayer('bmearns.c7').addTo(map);

      // add click handler for grid
      // changed variable name in tandem with change above
      c7Grid.on('click', function(e) {
        if (!e.data) return;
        var popup = L.popup()
          .setLatLng(e.latLng)
          // changed to refer to a field we have here, as seen in
            tilemill interaction tab
          .setContent(e.data.id)
          .openOn(map);
```

```
        });
      </script>
    </body>
  </html>
```

Previewing a simple UTFGrid modification

The preceding code will produce the following map view and grid interaction. Note that the value of the id attribute is displayed on click:

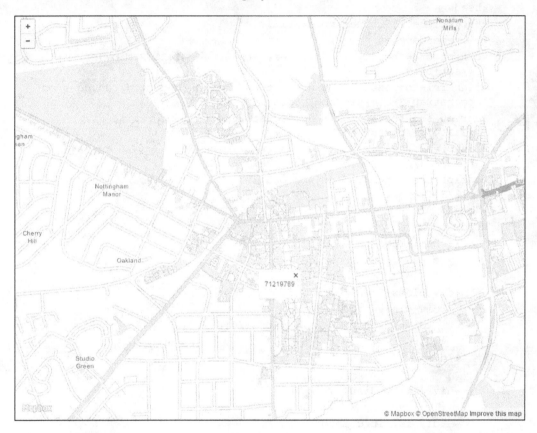

OpenLayers

The OpenLayers project provides a sample UTFGrid web map at http://openlayers.org/en/v3.2.1/examples/tileutfgrid.html.

Code modification

The following is the sample code with the modifications for our data. References to the OpenLayers example (for instance, in the function names) were removed and replaced with generic names. The following example demonstrates a mouseover type event trigger (`c7/data/web/utfgrid-ol.html`):

```html
<!DOCTYPE html>
<html>
  <head>
    <title>TileUTFGrid example</title>
    <!— dependencies -->
    <script src="https://code.jquery.com/jquery-1.11.2.min.js">
      </script>
    <link rel="stylesheet" href="https://maxcdn.bootstrapcdn.com/
      bootstrap/3.3.4/css/bootstrap.min.css">
    <script src="https://maxcdn.bootstrapcdn.com/bootstrap
      /3.3.4/js/bootstrap.min.js"></script>
    <link rel="stylesheet" href="https://cdnjs.cloudflare.com/ajax/
      libs/ol3/3.6.0/ol.css" type="text/css">
    <script src="https://cdnjs.cloudflare.com/ajax/libs/
      ol3/3.6.0/ol.js"></script>
  </head>
  <body>
    <!-- html layout -->
    <div class="container-fluid">

      <div class="row-fluid">
        <div class="span12">
          <div id="map" class="map"></div>
        </div>
      </div>

      <div style="display: none;">
        <!-- Overlay with target info -->
        <div id="info-info">
          <div id="info-name"> </div>
        </div>
      </div>

    </div>
    <script>
      // new openlayers tile object, pointing to TileJSON object
        from our mapbox service
      var mapLayer = new ol.layer.Tile({
```

```
      source: new ol.source.TileJSON({
        url: 'http://api.tiles.mapbox.com/v3/bmearns.c7.json?
          access_token=YOURMAPBOXTOKENHERE'
      })
    });
    // new openlayers UTFGrid object
    var gridSource = new ol.source.TileUTFGrid({
      url: 'http://api.tiles.mapbox.com/v3/bmearns.c7.json?
        access_token=YOURMAPBOXTOKENHERE'
    });

    var gridLayer = new ol.layer.Tile({source: gridSource});

    var view = new ol.View({
      center: [-8432793.2,4846930.4],
      zoom: 15
    });

    var mapElement = document.getElementById('map');
    var map = new ol.Map({
      layers: [mapLayer, gridLayer],
      target: mapElement,
      view: view
    });

    var infoElement = document.getElementById('info-
      info');
    var nameElement = document.getElementById('info-name');

    var infoOverlay = new ol.Overlay({
      element: infoElement,
      offset: [15, 15],
      stopEvent: false
    });
    map.addOverlay(infoOverlay);

    // creating function to register as event handler, to
      display info based on coordinate and view resolution
    var displayInfo = function(coordinate) {
      var viewResolution = /** @type {number} */
        (view.getResolution());
      gridSource.forDataAtCoordinateAndResolution(coordinate,
        viewResolution,
      function(data) {
```

```
          // If you want to use the template from the TileJSON,
          //  load the mustache.js library separately and call
          //  info.innerHTML = Mustache.render(gridSource.
            getTemplate(), data);
          mapElement.style.cursor = data ? 'pointer' : '';
          if (data) {
            nameElement.innerHTML = data['id'];
          }
          infoOverlay.setPosition(data ? coordinate : undefined);
        });
      };

      // registering event handlers
      map.on('pointermove', function(evt) {
        if (evt.dragging) {
          return;
        }
        var coordinate = map.getEventCoordinate(evt.original
          Event);
        displayInfo(coordinate);
      });

      map.on('click', function(evt) {
        displayInfo(evt.coordinate);
      });
    </script>
```

Putting it all together

Now, we'll connect the Google Sheets feed with our Mapbox tiles service in our final application.

Parsing the sheets JSON feed

Previously, we parsed the JSON feed with jQuery for each loop to print two attribute values for each element. Now, we'll remap the feed onto an object that we can use to look up data for the geographic objects triggered in UTFGrid. Review the following code, to see how this done:

```
// Create a data object in public scope to use for mapping
// of JSON data, using id for key
var d = {};

// url variable is set with code from previous section
```

```
// url contains public sheet id

$.getJSON(url, function(data) {
  var entry = data.feed.entry;
  var title = '';

  $(entry).each(function(index, value){
    // Column names are name, type, etc.
    $('.results').prepend('<h2>'+this.gsx$name.$t+'</h2><p>'+
      this.title.$t +'</p>'+'<p>'+this.gsx$type.$t+'</p>');
    title = this.title.$t;
    $.each(this, function(i, n){
      if(!d[title]){
        d[title] = {};
      }
      d[title][i] = n.$t;
    });
  });
});
```

Completing the application

Finally, to complete the application, we need to add the event handler function inside the jQuery AJAX call to the code which handles our feed. This will keep the mapped data variable in a scope relative to the events triggered by the user. The following code is in `c7/data/web/utfgrid-mb.html`:

```html
<!DOCTYPE html>
<html>
  <head>
  <meta charset=utf-8 />
  <title>A simple map</title>
  <meta name='viewport' content='initial-scale=1,maximum-
    scale=1,user-scalable=no' />
  <script src="http://code.jquery.com/jquery-1.11.3.
    min.js"></script>
  <script src='https://api.tiles.mapbox.com/mapbox.js/v2.2
    .1/mapbox.js'></script>
  <link href='https://api.tiles.mapbox.com/mapbox.js/v2.2
    .1/mapbox.css' rel='stylesheet' />
  <style>
    body { margin:0; padding:0; }
    #map { position:absolute; top:0; bottom:0; width:100%; }
  </style>
  </head>
  <body>
    <div id='map'></div>
```

```
<script>
// ID of the Google Spreadsheet
var spreadsheetID = "1gDPlmvEX0P4raMvTJzcVNT3JVhtL3e
  K1XjqE7u9u4W4";
// Mapbox ID
L.mapbox.accessToken = 'pk.eyJ1IjoiYm1lYXJucyIsImEiO
  iI1NTJhYWZjNmI5Y2IxNDM5M2M0N2M4NWQyMGQ5YzQyMiJ9.q8-
  B7BXtuizGRBcnpREeWw';

var map = L.mapbox.map('map', 'mapbox.streets')
  .setView([39.87240,-75.75367], 15);

c7tiles = L.mapbox.tileLayer('bmearns.c7').addTo(map);
c7grid = L.mapbox.gridLayer('bmearns.c7').addTo(map);

// Setup click handler, Google spreadsheet lookup
var d = {};
// Make sure it is public or set to Anyone with link can
  view
var url = "https://spreadsheets.google.com/feeds/list/" +
spreadsheetID + "/od6/public/values?alt=json";
$.getJSON(url, function(data) {

  var entry = data.feed.entry;
  var title = '';

  // loops through each sub object from the data feed using
    json each function, constructing html from the object
    properties for click handler
  $(entry).each(function(index, value){
    // Column names are name, type, etc.
    $('.results').prepend('<h2>'+this.gsx$name
      .$t+'</h2><p>'+
      this.title.$t +'</p>'+'<p>'+this.gsx$type.$t+'</p>');
    title = this.title.$t;
    $.each(this, function(i, n){
      if(!d[title]){
        d[title] = {};
      }
      d[title][i] = n.$t;
    });
  });
  // register click handler, displaying html constructed
    from object properties/loop
  c7grid.on('click', function(e) {
    if (!e.data) return;
    key = e.data.id;
    content = d[key].content
```

```
        //
    var popup = L.popup()
        .setLatLng(e.latLng)
        .setContent(content)
        .openOn(map);
    });
});


    </script>
  </body>
</html>
```

After saving this with a .html extension (for example, `c7/data/web/utfgrid-mb.html`), you can preview the application created using this code by opening the saved file in a web browser. When you do so, you will see something similar to the following image:

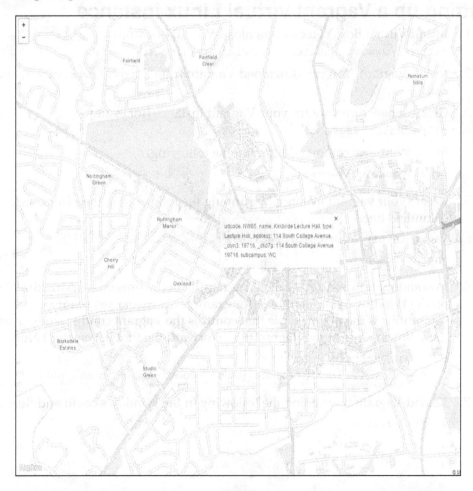

Going further – local MBTiles hosting with TileStream

While Mapbox hosting is very appealing for its relative ease, you may wish to host your own MBTiles, for example, to minimize cost. TileStream is the open source foundation of Mapbox's hosting service. It runs under Node.js.

While TileStream can technically be installed under Windows with a Node.js install, some dependencies may fail to be installed. It is recommended that you perform the following install in a Linux environment. If you are already running Linux on your organization's server, that's great! You can skip ahead to installing Node.js and TileStream. Fortunately, with Virtual Box and Vagrant, it is possible to set up a Linux virtual machine on your Windows system.

Setting up a Vagrant virtual Linux instance

1. Install Virtual Box. You can download Virtual Box from `https://www.virtualbox.org/wiki/Downloads`.

2. Install Vagrant. You can download Vagrant from `https://www.vagrantup.com/`.

3. Create a new directory for your Vagrant instance (for example, `c:\packt\c7\vagrant`).

4. In the Windows command line, run the following:

   `cd c:\packt\c7\vagrant`

5. Create your vagrant instance by running the following in the Windows command line:

   `vagrant init hashicorp/precise32`

   `vagrant up`

6. You should now see a file called Vagrantfile in the present working directory on the Windows command line (for example, `c:\packt\c7\vagrant`). Vagrantfile is the plaintext file that controls the Vagrant configuration. Add the following line to the Vagrantfile to forward the (not yet created) Node.js port to your Windows localhost:

   `config.vm.network "forwarded_port", guest: 8888, host: 8088`

7. Reload Vagrant by running the following in the Windows command line:

   `vagrant reload`

8. Connect to the Vagrant instance through a Windows SSH client.

   ° If you already have an SSH client on the PATH environment, the `vagrant ssh` variable should start it up and connect it to the instance in this directory

   ° If not, you can just set the following connection parameters under your Windows SSH client of choice, such as Putty:

   Host: `127.0.0.1`

   Port: `2222`

   Username: `vagrant`

   Password (if needed): `vagrant`

   Private key: `c:/packt/c7/vagrant/.vagrant/machines/default/virtualbox/private_key`

Installing Node.js and TileStream

Use the following commands to install Node.js from the `chris-lea` package archive. This is the Node.js source recommended by Mapbox. The preceding Vagrant setup steps ensure that the dependencies, such as `apt`, are set up as needed. Otherwise, you may install some other dependencies. Also, note that on systems that do not support GNU/Debian, this will not work; you will need to search for the relevant repositories on yum, find RPMs, or build it from the source. However, I'm not sure that any of this will work.

Note that all the following commands are to be run in the Linux instance. If you followed the preceding steps, you will run these through an SSH client such as Putty:

```
sudo apt-add-repository ppa:chris-lea/node.js
sudo apt-get update
sudo apt-get install nodejs
```

Install Git and clone the TileStream source to a new directory. Run the following command line:

```
git clone https://github.com/mapbox/tilestream.git
cd tilestream
npm install
```

Setting up and starting TileStream

You must add your MBTiles file under `/home/vagrant/Documents/MapBox/tiles`. You can use your preferred file transfer client to do this; I like WinSCP. Just use the same SSH connection info you used for SSH.

Finally, start TileStream:

```
./index.js
```

Now, you can preview your TileStream service at `http://localhost:8088`.

Click on the info button to obtain the address to the PNG tiles. You can modify this by removing the reference to the *x*, *y*, and *z* tiles and adding a `.json` extension to get TileJSON such as `http://localhost:8088/v2/c7_8b4e46.json`.

Now, simply modify the OpenLayers example to refer to this `.json` address instead of Mapbox, and you will have a fully nonMapbox use for MBTiles.

The code demonstrating this is at `http://localhost:8000/utfgrid-ts.html`.

Summary

In this final chapter, we looked at a web application built on the web services that provides editing capabilities to our user. This is on the simpler end of collaborative geographic data systems but with an attractive cartographic rendering capability offered by TileMill (and Mapnik) and a highly performant data publishing capability through MBTiles and UTFGrid.

Index

W

web application
  about 121, 180
  API access 121
  testing 190
  Twitter account, registering 121
  Twitter Tools API, setting up 122-125
web mapshaper
  about 147
  URL 147
web publishing 3
Well Known Text (WKT) 3

Z

zonal statistics 54-7

Thank you for buying
QGIS Blueprints

About Packt Publishing

Packt, pronounced 'packed', published its first book, *Mastering phpMyAdmin for Effective MySQL Management*, in April 2004, and subsequently continued to specialize in publishing highly focused books on specific technologies and solutions.

Our books and publications share the experiences of your fellow IT professionals in adapting and customizing today's systems, applications, and frameworks. Our solution-based books give you the knowledge and power to customize the software and technologies you're using to get the job done. Packt books are more specific and less general than the IT books you have seen in the past. Our unique business model allows us to bring you more focused information, giving you more of what you need to know, and less of what you don't.

Packt is a modern yet unique publishing company that focuses on producing quality, cutting-edge books for communities of developers, administrators, and newbies alike. For more information, please visit our website at www.packtpub.com.

About Packt Open Source

In 2010, Packt launched two new brands, Packt Open Source and Packt Enterprise, in order to continue its focus on specialization. This book is part of the Packt Open Source brand, home to books published on software built around open source licenses, and offering information to anybody from advanced developers to budding web designers. The Open Source brand also runs Packt's Open Source Royalty Scheme, by which Packt gives a royalty to each open source project about whose software a book is sold.

Writing for Packt

We welcome all inquiries from people who are interested in authoring. Book proposals should be sent to author@packtpub.com. If your book idea is still at an early stage and you would like to discuss it first before writing a formal book proposal, then please contact us; one of our commissioning editors will get in touch with you.

We're not just looking for published authors; if you have strong technical skills but no writing experience, our experienced editors can help you develop a writing career, or simply get some additional reward for your expertise.

Learning QGIS 2.0

ISBN: 978-1-78216-748-8   Paperback: 110 pages

Use QGIS to create great maps and perform all the geoprocessing tasks you need

1. Load and visualize vector and raster data.

2. Create and edit spatial data and perform spatial analysis.

3. Construct great maps and print them.

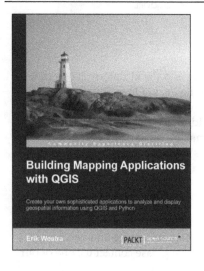

Building Mapping Applications with QGIS

ISBN: 978-1-78398-466-4   Paperback: 264 pages

Create your own sophisticated applications to analyze and display geospatial information using QGIS and Python

1. Make use of the geospatial capabilities of QGIS within your Python programs.

2. Build complete standalone mapping applications based on QGIS and Python.

3. Use QGIS as a Python geospatial development environment.

Please check **www.PacktPub.com** for information on our titles

Learning QGIS

Second Edition

ISBN: 978-1-78439-203-1   Paperback: 150 pages

Use QGIS to create great maps and perform all the geoprocessing tasks you need

1. Load, visualize, and edit vector and raster data.

2. Create professional maps and applications to present geospatial data.

3. A concise guide, packed with detailed real-world examples to get you started with QGIS.

Mastering QGIS

ISBN: 978-1-78439-868-2   Paperback: 420 pages

Go beyond the basics and unleash the full power of QGIS with practical, step-by-step examples

1. Learn how to meet all your GIS needs with the leading open source GIS.

2. Master QGIS by learning about database integration, geoprocessing tools, Python scripts, advanced cartography, and custom plugins.

3. Create sophisticated analyses and maps with illustrated step-by-step examples.

Please check **www.PacktPub.com** for information on our titles